INCIDENT COMMAND
FOR
EMS

INCIDENT COMMAND
FOR
EMS

Karen Owens

Fire Engineering

Disclaimer

The recommendations, advice, descriptions, and methods in this book are presented solely for educational purposes. The author and publisher assume no liability whatsoever for any loss or damage that results from the use of any of the material in this book. Use of the material in this book is solely at the risk of the user.

Copyright © 2012 by
PennWell Corporation
1421 South Sheridan Road
Tulsa, Oklahoma 74112-6600 USA

800.752.9764
+1.918.831.9421
sales@pennwell.com
www.Fire EngineeringBooks.com
www.pennwellbooks.com
www.pennwell.com

Marketing Coordinator: Jane Green
National Account Executive: Cindy J. Huse

Director: Mary McGee
Managing Editor: Marla Patterson
Production Manager: Sheila Brock
Production Editor: Tony Quinn

Library of Congress Cataloging-in-Publication Data

Owens, Karen, 1980–
    Incident command for EMS / Karen Owens.
        p. ; cm.
    Includes bibliographical references and index.
    ISBN 978-1-59370-267-0
    I. Title.
    [DNLM: 1.   Emergency Medical Services--organization & administration--United States. 2.   Disaster Planning--organization & administration--United States.   WX 215]

    LC classification not assigned
    362.18068--dc23
                                2011031328

Printed in the United States of America

1   2   3   4   5   16   15   14   13   12

I would like to dedicate this book to my husband, Robby, and my two sons, Junior and Ben, who spent many nights watching me sit at the computer typing away while they played. I cannot thank you enough for the support, encouragement, and occasional motivational kick that you all provided me to get this finished. Without you, I might not have accepted the opportunity to write this book. Your belief that I could do it was an important factor.

I would also like to thank my family and my in-laws for the support they have provided. From proofreading to baby sitting, this book would have never been completed without you. And while my mom will never get to see the final version, I know that I had her love and support. Your support through this process has been amazing. To my friends who have supported and encouraged me, thank you.

I would like to thank my boss, Jim Nogle, and the director of the Virginia Office of EMS, Gary Brown. Their support as I took on this project, as well as the opportunities they provide me on a daily basis, allowed me to gather the necessary research to ensure that this book met the needs of the EMS providers.

I would also like to thank my editor and the publisher for providing me with this amazing opportunity. I am humbled and honored!

And to everyone who is running the calls and helping the community, thank you for continuing to do your job and training to do it even better.

CONTENTS

THE BACKGROUND AND BASICS OF INCIDENT COMMAND SYSTEMS

Chapter Objectives

Upon completion of this chapter, readers will:

1. Know the general history of the incident command system (ICS)

2. Recognize federal regulations and standards that impact the use of ICS at emergency scenes

3. Understand the basic principles of the ICS design

4. Recognize the components of the ICS structure

5. Recognize and understand the roles of the five major components of the ICS

The Background of ICS

The 1970s saw a significant increase in the number of wildfires that threatened the residents of California. In 1970, a series of wildfires claimed 16 lives, burned down at least 700 structures, and scorched more than a half million acres of land.[1] In a review of the incident response, it was agreed that while participating agencies cooperated to the best of their abilities, response was made less effective due to communications problems between the agencies. In 1971, as a result of the significant loss of life and property, Congress provided funding to the U.S. Forest Service to strengthen fire command and control systems research at Riverside, California, and Fort Collins, Colorado. The intent of the research program was to design a system that would "make a quantum jump in the capabilities of Southern California wildland fire protection agencies to effectively coordinate interagency action and to allocate suppression resources in dynamic, multiple-fire situations."[2] The system that was designed by the Forest Service became known as FIrefighting REsources of Southern California Organized for Potential Emergencies, or FIRESCOPE. As FIRESCOPE grew, additional agencies joined the movement including the California Department of Forestry and Fire Protection; the Governor's Office of Emergency Services; the Los Angeles, Ventura, and Santa Barbara County fire departments; and the Los Angeles Fire Department. These agencies put the practices and methods of the FIRESCOPE system into place and began to utilize them.

In the efforts to develop a system, four requirements became obvious to group members.[3] They were:

1. The system must be organizationally flexible to meet the needs of incidents of any kind and size.

2. Agencies must be able to use the system on a day-to-day basis for routine situations as well as for major emergencies.

3. The system must be sufficiently standard to allow personnel from a variety of agencies and diverse geographic locations to rapidly meld into a common management structure.

4. The system must be cost effective.

These four requirements became the building blocks of the first basic incident command system structure.

Federal government support

Regulations affecting the response to hazardous materials incidents were some of the first written requirements used to manage incidents in a coordinated fashion. The first regulation, from the Superfund Amendments and Reauthorization Act (SARA), was Title III. Title III states that the Occupational Safety and Health Administration (OSHA) is responsible for creating and implementing rules and regulations for operations at a hazardous materials incident. Title III also protects a community's right to know what hazardous materials are maintained in their community so that they have the ability to prepare for a potential response. With the push from SARA Title III, OSHA created additional regulations that govern incident coordination and response. OSHA 29 CFR Part 1910.20 requires all organizations—fire, EMS, police, and other—that respond to hazardous materials incidents use an incident command system structure during response. It further states that those employers who create an ICS structure should ensure that it is compatible with agencies that may respond to a hazardous materials incident within their business. For those states that are not required to follow OSHA standards, the Environmental Protection Agency (EPA) has established requirements for the use of ICS at hazardous materials incidents (40 CFR Part 311).[4]

In an effort to help solidify the federal government's support of the push for a national incident management system, President George W. Bush signed two Presidential directives. The first, Homeland Security Presidential Directive 5 (HSPD-5), is designed to "enhance the ability of the United States to manage domestic incidents by establishing a single, comprehensive national incident management system."[5] Signed February 28, 2003, this presidential directive was the beginning of the national standard for an incident management structure that allows for a smoother integration of resources during emergency situations. Because of the directive, steps were taken to create a standardized system that became a requirement for all agencies, fire and rescue included, that participate in disaster response. By ensuring that every responder has met a standardized set of requirements, including training and practice, the hope is that mutual aid and disaster response will be more efficient and effective.

The National Fire Protection Association (NFPA) standards supporting the use of ICS during emergency response include:

- **NFPA 1500** – requires that all agencies that have adopted the standard establish written procedures for an ICS and train all members in ICS.

- **NFPA 1561** – establishes broad guidelines for what should be included in an ICS.

- **NFPA 1600** – establishes a minimum standard for disaster management.

- **NFPA 1710** – establishes minimum requirements relating to the organization and deployment of fire suppression and emergency medical operations to the public.

- **NFPA 1720** – establishes the minimum requirements relating to the organization and deployment of fire suppression resources; and for those fire departments that provide them, emergency medical and special operations resources.

- **NFPA 473** – identifies the level of competence required by EMS personnel who respond to hazardous materials incidents.

A second directive signed by President Bush, HSPD-8, was written as a companion document to HSPD-5. Signed December 17, 2003, HPSD-8 establishes improvements to the delivery of federal assistance provided to states and local governments and requires an all-hazards preparedness goal.[6] The directive also outlines actions to strengthen preparedness capabilities of federal, state, and local entities. This directive focuses on response efforts of the federal government, working to ensure that the assistance offered to state and local governments is done so in a timely and effective manner, and that the preparedness of federal resources is for all potential incidents, not just a small set.

From ICS to NIMS

While ICS is a standardized approach to incident response, the national incident management system (NIMS) provides a more comprehensive approach to emergency response through all four phases of emergency management: mitigation, preparedness, response, and recovery. NIMS combines the best practices of emergency response and emergency management to create a systematic approach to emergency response management. In recognition of its benefit to emergency response, the national incident management system utilizes ICS as one piece of the entire system.

After the release of HSPD-5 with a directive to create the national incident management system, meetings were conducted with representatives from state, local, and federal government, along with private and public safety organizations. On March 1, 2004, as a result of these meetings, the structure and guidelines for NIMS implementation were released.[7] NIMS provides a common ground for all responders, from fire and EMS to public health and education workers. It provides principles, terminology, and processes to support effective incident command operations. Unlike previous variations of the ICS, NIMS ICS focuses on the impact of the larger incident response on the ICS structure. Recognition is given to the fact that law enforcement, public health, and federal government agencies, as well as other private and public partners, will participate in the ICS structure when an incident is large enough and changes are made to help with this integration.

NIMS ICS and FIRESCOPE ICS are similar in the duties and responsibilities of staffed positions. Common terminology can be found in the position titles of staff and also in the layout of the ICS structure. However, one glaring difference between NIMS ICS and FIRESCOPE ICS is the creation of an information and intelligence function. The NIMS ICS structure creates a sixth functional area that, when necessary, coordinates analysis and sharing of information and intelligence. This information may include but is not limited to risk assessments, medical intelligence, weather information, and building design and hazard information.

The NIMS structure focuses on five major components in its approach to incident management. The first component is preparedness. The creators of the NIMS structure recognize the need to be appropriately prepared for all incidents that might occur in a locality. NIMS emphasizes an integration of planning, procedures and protocol, training and exercise, personnel qualifications and certification, and equipment certification to assist localities in preparedness measures.

The second component of NIMS is communications and information management. With this component, NIMS defines a standardized framework of communications, basing the framework on interoperability, reliability, scalability, portability, and redundancy. Each of these characteristics is necessary to ensure effective and efficient communications during incident response.

The third major component of the NIMS system is resource management. The NIMS structure recognizes that people, equipment, and supplies must flow steadily to and from an incident for operations to run successfully. NIMS creates a mechanism for resource identification, typing, mobilizing, reimbursement, and inventory.

Command and management is the fourth component of the NIMS system. This component ensures that incident management is effective and efficient through the utilization of flexible structures that surround the constructs of incident command systems, multia-gency coordination systems, and public information.

The final component of NIMS is ongoing management and maintenance. The Department of Homeland Security and the Federal Emergency Management Agency follow processes for maintenance of NIMS training programs and updates to ensure that the NIMS structure meets the current needs of response agencies.

To assist in the integration and oversight of NIMS program implementation, the National Integration Center (NIC) was created by the secretary of Homeland Security. The NIC is tasked with overseeing development, management, and coordination of training, education, and exercise of topics related to homeland security, including incident command systems training, to assist in emergency response to all incidents whether natural or manmade.[8] This also includes review of lessons learned from exercises and events in which NIMS is utilized. The NIC divides its activities into different branches to allow for focus on many different integration aspects. These branches include standards and resources; training and exercises; system evaluation and compliance; technology, research, and development; and publications management.[9] This allows for efficient assistance to localities and states as they work through the implementation process.

In support of efforts to implement NIMS at every level, the NIC suggests steps to jurisdictions. These steps are:

1. Incorporate NIMS into existing training programs and exercises.

2. Ensure federal preparedness funding supports implementation programs.

3. Incorporate NIMS into emergency response plans.

4. Promote mutual aid agreements within intrastate agencies.

5. Coordinate technical assistance between state and local entities.

6. Institutionalize the use of an emergency operations center (EOC) in operations.

Each of these suggested steps allows for support as the process of implementation moves forward. This support can be provided both at the operational level and the administrative level. Knowing that there is support for implementation allows you to focus on training your personnel and coworkers on the basics of the system.

The Basics of ICS

The ICS structure is a standardized all-hazard approach to incident response. It provides an integrated response to all situations faced by emergency response personnel at all types of incidents. There are multiple benefits to the use of this structure. First, it meets the needs of incidents of any kind or size. Regardless of the number of individuals responding, the area of the incident, or the extent of the damage that occurred, ICS can work. A second benefit is that the use of ICS ensures that personnel from a variety of agencies are able to work together in a common command structure so that no time is lost in training personnel to different response structures. A third benefit is that following the ICS structure ensures that logistical and administrative support is provided to the operational functions at the scene. The logistical and administrative support allows for resources, equipment, and paperwork to be completed in a timely manner and allows operational personnel to focus on the functional aspects of response. The final benefit of the ICS structure is that it is a cost effective method of coordinating response.

Standard characteristics

One characteristic found in the ICS structure is management by objectives. Under this principle, objectives set by incident command are based on feedback from you and your coworkers. Objectives are set based on information gathered from all levels of the organization after review by the command staff. This ensures that assignments made are based on realistic goals and that subordinates will have a greater motivation to complete their assignments based on their personal input.

Unit integrity is another characteristic practiced in an effective ICS structure. Under this principle, roles defined for EMS providers make sense organizationally and strategically. For instance, EMS providers are not assigned to police or public works duties, and police officers are not assigned to firefighting tasks. Not only does this ensure that the work is being done by the appropriate person, but it also helps with liability and record keeping throughout the incident.

Functional clarity is the third characteristic of the ICS structure. Under this concept, each part of the organization is developed to ensure that the focus is appropriately directed. Tasks and responsibilities of each operational worker are clearly defined, and when one task is accomplished, additional tasks are given to ensure that every person on scene has a clear purpose.

Additionally, span of control is another characteristic associated with an effective ICS structure. In order to ensure effective and efficient operations, each leader/supervisor should have as few as three and as many as seven direct reporting subordinates. In instances where each leader has only a few subordinates, you create a structure with too many leaders and not enough operational staff. In an incident where each leader has a large number of subordinates, you create a structure where operations are less efficient because there are not enough leaders to make decisions.

Another characteristic associated with a successful ICS structure is chain of command. This principle refers to effective and efficient information flow within the structure, both up (from the field to the leaders) and down (from the leaders to the field). When designing an ICS structure for an incident, you should keep in mind that ICS is based on modular organization, a third characteristic of ICS. This means that functional assignments are grouped together so that the ICS structure can expand and contract to meet the needs of an incident. A simple incident may not require the resources associated with a certain branch of ICS and because of this, that branch does not need to be staffed during the incident. A modular organization ensures that the size of the structure is appropriate for the size of the incident.

A final characteristic, and one some consider the most valuable, is common terminology. The design of the incident command system is intended to allow for the integration of a variety of agencies during response to both small and large incidents. In working to ensure a smooth integration of different agencies, the need for common terminology was recognized. Common terminology extends to many aspects of ICS, including organizational positions, facilities, and resource elements.

Organizational positions. By using common terminology to name the organizational positions within ICS, you can assign roles without having to explain the responsibilities and duties of the role, regardless of the jurisdiction of a responder. Using the appropriate terminology ensures a smooth transition in multi-agency response. Positions are titled based on their organizational level. Table 1–1 shows the titles and associated support positions found in the ICS.

Table 1–1 Titles given to various positions within the ICS structure and support positions.[10]

Organizational Level	Title	Support Position
Incident command	Incident	Deputy
Command staff	Officer	Assistant
General staff (section)	Chief	Deputy
Branch	Director	Deputy
Division/group	Supervisor	N/A
Unit	Leader	Manager
Area	Manager	
Strike team/task force	Leader	Single resource boss

Facilities. From the simple to the complex, emergency incidents require the need of support facilities to ensure effective operations. As an EMS provider at an incident, you will rely on the functioning of many of these facilities to assist in the treatment and transport of patients and injured on-site personnel. The following is a list of facilities and their functions:

- Incident command post (ICP). The incident command post is the location from which the incident commander (IC) or unified command structure manages the incident. There is only one ICP for each incident and it should be located just outside of the incident, still providing a good view of the operations (figure 1–1). Once established, the IC should announce the locations of the ICP and ensure that it is appropriately marked.

Fig. 1–1 The incident command post (ICP) should provide space for all members of command to gather to make operational decisions for the incident.

- Staging area. The location where all incoming resources first report until they are assigned a specific duty on the incident scene is called the staging area. This area should be large enough to accommodate all associated resources, from ambulances to tow trucks or cranes, that may be necessary to handle the incident. The staging area is supervised by the staging area manager, who reports directly to the operations section chief.

- Base. The base is the location where primary incident response activities occur. There may not be a base established at incident, but when established, there will only be one base.

- Camp. Similar to a staging area, a camp is an area where resources go to assist in incident operations. However, unlike resources in the staging area, resources at a camp may not be readily available for duty assignment.

- Helispot. Under NIMS concepts, the location often referred to as the landing zone (LZ) is now known as a helispot. The helispot is the area where an aeromedical unit may land for the loading and offloading of patients, people, equipment, and/or supplies. The location of the helispot is important because it can impact the transportation of patients from the treatment area to the aeromedical unit. The treatment area is discussed further in chapter 3.

- Helibase. The helibase is the location where multiple aeromedical units can land to refuel, conduct maintenance, or resupply equipment. It should be able to handle more than one unit. There can be multiple helibases at an incident.

Resource elements. Common terminology when naming resources is important, especially when asking for assistance. Ensuring that all agencies utilize common names for equipment helps clarify resource requests.

The Components of ICS

When appropriately set up, the ICS structure can integrate over 5,000 personnel. By creating a strong basic structure you can ensure that an operation is quickly brought under control and patient care is provided in a timely and effective manner. The basic structure of ICS includes the following positions: incident commander, operations, logistics, planning, and finance/administration. This basic structure is applied to any incident (non-emergency or emergency) and is consistent for every incident. The five basic components provide a strong foundation for the rest of the structure.[11]

General Staff

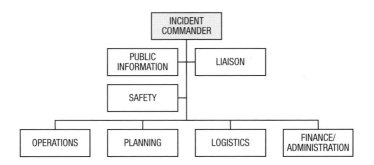

Incident commander

Based on the ICS standards, the first responder to arrive at an incident scene staffs the incident command position until someone arrives that can take the position. Because the success of the ICS structure relies on appropriate implementation from the start, the first role filled is that of incident commander (IC).

Who Is in Charge?

Within your agency, you know who your supervisors are and who the head of the agency is, but that may not translate to the incident command system structure in the field. For many, the question of who is in charge at the incident is not an easily answered question. However, a few regulations may help answer the question. Under OSHA 29 CFR 1910.120, paragraph Q states that the "senior emergency response official responding to an emergency shall become the individual in charge of a site-specific Incident Command System (ICS)."[12] The OSHA standard further defines the senior official as the individual who has most seniority. As senior officers arrive (fire chief, battalion chief, etc.), the position is passed to those individuals.

Additionally, some states have written regulations that set the standard for the individual who will hold the role of incident commander. While most of these regulations state that the EMS leadership on scene are in charge of patient care aspects of the incident, there may come a point when EMS providers are looked at to hold the incident command position, even if for a short period of time. With this in mind, you should be familiar with the roles and responsibilities of all positions in the incident command system structure.

The individual staffing the incident command position is referred to as the incident commander. The IC is responsible for coordination of activities within the incident command system. This person maintains the overall "big picture" at the incident and does not become directly involved in operational aspects of the incident, instead ensuring that all incident objectives are met. The IC is responsible for the duties of any of the five positions of the basic structure that are not staffed as well as the duties of that position. From the beginning of the incident to the end, the IC holds ultimate responsibility for the entire incident, whether or not he or she has direct contact with the personnel. The following are the responsibilities of the incident commander.

Assess situation. From the initial size-up to continued review of the incident, the IC must carry out the assessment of the situation. This may be done upon initial establishment of the incident command function or through a briefing from the outgoing IC.

Determine incident objectives and strategy. With feedback and input from all personnel on the scene (done so through an appropriate chain of command), the IC is responsible for setting and disseminating the incident objectives. These objectives dictate the actions taken by on-scene personnel and also the resources requested to ensure mitigation of the event.

Establish immediate priorities. The immediate priorities of an incident will not change, regardless of the type of incident. They are life safety, incident stabilization, and property conservation. From your first training at the EMT-basic level, you have learned that your safety, the safety of your crew, and the safety of bystanders and the patient are always priority. That has not changed. With an ever-changing scene, the IC must ensure that the immediate priorities are controlled and take steps to ensure that they continue to be met throughout the incident. Maintaining these priorities can also change based on available resources and incident conditions.

Establish an incident command post (ICP). Most utilized at large and complex incidents, an incident command post provides a central location for command personnel to gather and make decisions. When determining where to establish the command post, the IC should consider the ability to view as much of the incident as possible while in the command post. The IC should also ensure that the ICP is in a safe location and that it minimally impacts the tactical activities of the operation. Once established, the ICP should be clearly marked so that it is visible to on-scene personnel.

Establish an appropriate organization. As previously mentioned, one major benefit of the ICS structure is that it is modular and can expand or contract based on the incident needs. It is the role of the IC to determine who should serve in which functions based on their knowledge, skills, abilities, and experience. The IC must also determine whether the incident will best be managed by a single incident commander, unified command (figure 1–2), or area command.

Single incident command. This type of leadership is standard for most incidents, with a single individual carrying out the functions of the incident commander position.

Unified command. In a more complex incident, a unified command structure is used to oversee an incident. Unified command relies on representatives from all agencies to assist in setting tactics and strategies.

Area command. An area command is used to manage a very large incident that has multiple incident management team assignments. It can be used when incident scenes are close enough that conflict may arise if coordination is not carried out.

Approve and authorize the incident action plan. In order to have successful coordination of activities, every incident must have an incident action plan (IAP), either written or verbal. An incident action plan is written for an operational period and shares with the on-scene personnel an understanding of the incident objectives and a clear direction on how to meet the objectives. A new IAP should be written for each operational period and changed based on the completion of objectives and changes in the incident.

Fig. 1–2 A unified command structure brings together representatives from all response agencies, both private and public.

Ensure safety of personnel. Safety is the responsibility of every person at an incident scene; however, it is the ultimate responsibility of the IC to ensure that safety measures are in place to protect all workers. At larger and more complex incidents, safety is normally of greater concern. As needed, the IC can appoint a safety officer to focus specifically on the safety needs of the incident.

Coordinate activity for all command and general staff. Along with maintaining situational awareness, the IC is responsible for the coordination of activity between command and general staff. In order to ensure this coordination, the IC should schedule periodic briefings from staff members, maintain good communication (preferably face-to-face), and conduct planning meetings. This maintains a standard level of awareness and understanding among all staff members.

Coordinate with key personnel and officials. Along with the on-scene personnel, the incident commander may find a need to interact with key personnel and officials from the agency and/or jurisdiction. This interaction should not take precedence over incident mitigation. Coordination efforts may, however, be necessary to ensure the support of off-site personnel in management positions.

Approve requests for additional resources or release of resources. When it is determined that additional resources are necessary to effectively handle an incident, it is ultimately the responsibility of the IC to make the request. After feedback from general staff members, the IC may make the request through the appropriate process. The reverse is also true. When resources are no longer needed, the IC should release the resources to return to their home agency or jurisdiction.

Authorize release of information to media. While the position of the public information officer can be staffed within the ICS structure, approval for the release of any incident related information must be received from the IC. This authorization is harder to control at larger/more complex incidents; however, it is important to ensure continuity of information.

Order the demobilization of the incident when appropriate. The IC must recognize that an incident is coming to a close and take measures to ensure that the demobilization of resources is done in a controlled manner. Larger incidents may require a written plan, while smaller incidents may be able utilize a verbal demobilization plan. One demobilization decision for the IC to make is whether to send first on-scene resources home first or to send outside agency resources home first.

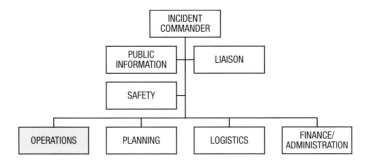

Operations

In most applications of the ICS structure, the operations section is the second function staffed. Operations is a very active function, as its main charge is to carry out tasks decided upon by the incident commander and placed in the incident action plan.

Operations section chief. As with other sections, there is a supervisor of the operations section. The individual who serves as the supervisor of this section is the operations section chief. The primary responsibility of the operations section chief is the management of resources to ensure successful completion of tactical assignments. Just like the IC, the operations section chief has a set of duties and responsibilities that he or she must carry out throughout the incident. The following are the responsibilities of the operations section chief.

Develop operations portion of the incident action plan. The operations section chief must coordinate with the planning section to develop the operations portion of the IAP. This includes determining the tactics that will be used and resources necessary to meet the incident objectives.

Brief and assign operations section personnel in accordance with the IAP. After the operations section chief has an understanding of tactical assignments from the IC, it is important that the personnel staffing the subordinate positions within the operations section are briefed on the expectations. When appropriate, information flow should follow the appropriate chain of command.

Supervise operations section. One of the most important functions of the operations section chief is the supervision of all activities of the operations section. He or she must maintain an overall view of the actions of the operations section and ensure that interaction with other sections of the ICS structure is effective. Information obtained while supervising the activities of the operations section should be provided to the incident commander and the planning section so changes to the IAP can be made.

Determine needs and request additional resources. Once the operations section chief is briefed on the IAP, it is up to him or her to determine the short-term and long-term resource needs to meet the incident objectives. Once the needs are determined, the operations section chief must work with the incident commander, planning section, and logistics to request necessary resources. Consideration should be given to length of operation, response delay of resources, and other factors that may impact the number of resources needed. Also, once the operations section chief has requested additional resources, a staging area must be established to locate resources until they are needed in the operational area.

Assist in the determination of resources to be demobilized. While the planning section maintains the master roster of all on-scene resources, the operations section chief assists the IC in the decision of which resources can demobilize. Based on the demobilization of resources, the operations section chief may also make changes.

Operations support staff

There are multiple positions under the operations section chief that EMS providers will interact with in order to complete their assigned tasks. These positions, while not always staffed, serve as cooperative staff in working an incident.

Staging area manager. The staging area manager is responsible for oversight of the staging area. The staging area should be established quickly to ensure that responding units do not all congregate at the incident site. The staging area manager position is one that may be staffed based on availability of personnel rather than a particular assignment, thus all responders should be familiar with those responsibilities. The staging area manager must determine the layout of the staging area. Keep in mind that not only are ambulances directed to staging, but so are fire apparatus, as well other public and private resources. The staging area manager must also ensure that staging resources are checked in and checked out as appropriate based on assignments. This allows for appropriate accountability of resources and an accurate record of assigned versus available resources. The staging area manager is also responsible for ensuring that appropriate levels (as predetermined by the incident commander) of resources are maintained in the staging area. When levels reach a minimum, additional resources may be dispatched to assist the incident. From the EMS perspective, the staging area manager will work with the appropriate incident personnel to ensure appropriate transport resources are available and assigned as needed to best benefit any patients.

Air operations branch director. Use of aeromedical transportation units may be necessary for effective and efficient movement of patients from the incident site. When any helicopter or air unit is dispatched to an incident scene, the air operations branch director position is staffed. Tasks of the AOBD include development of the air operations branch, coordination between air units and ground personnel, and providing overall logistical support to the helicopters. Personnel within the air operations branch support EMS operations by coordinating the delivery of patients to aeromedical units for transport to medical facilities.

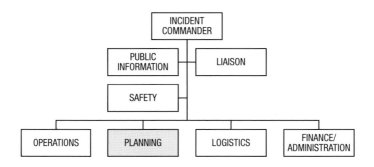

Planning

The planning section functions as a support to the operations and command functions within the ICS structure. The planning section is most often utilized during large and complex incidents, as the roles of this section are most necessary at those incidents. It is the decision of the incident commander to determine whether to implement the planning section. This decision is normally based on the complexity of the incident and whether the IC will need assistance in developing and releasing pertinent guides and plans regarding the incident. The planning section is led by the planning section chief. The ideal individual chosen to fill this role has knowledge, skills, and abilities associated with the job functions of the position. These job functions include:

Collect and process situation information. Pertinent information is gathered from all levels within the ICS structure at an incident. Reports are produced during each operational period and must then be analyzed so the information can be used by the IC to make changes to the IAP, consider utilizing new tactics, or prepare for demobilization of the incident.

Supervise preparation of the incident action plan and provide input as necessary. While it is the responsibility of each section chief to provide input for their section into the IAP, the planning section chief is responsible for ensuring that the IAP is developed using the appropriate guidelines and forms. Because the planning section has information from each section, the planning section chief may also serve as an advisor to the IC and operations section chief to provide coordination for effective scene management and provide alternate suggestions on an as-needed basis.

THE INCIDENT ACTION PLAN (IAP)

The IAP reflects the overall strategy that personnel should focus on during each operational period. Every section in the ICS structure is responsible for providing input in the development of the IAP, but the planning section is responsible for its overall development. While every plan is different, there are four elements that should be included in each plan:

- What do we want to do and how are we going to do it?

- Who is responsible for doing it?

- What is our communications plan?

- How will we handle injured personnel?

Each of these questions should be answered and those answers provided to the personnel carrying out the tactics. The best way to provide personnel with the information is through written documentation using various ICS forms developed by FEMA.[13]

Reassign out-of-service personnel to organizational positions as appropriate. As the planning section chief recognizes the need, he or she uses non-assigned personnel to staff organizational positions. This increases the efficiency of on-scene operations and ensures that span of control is maintained.

Establish information requirements and reporting schedules. In order to ensure that information flows consistently throughout the incident, the planning section chief must develop a schedule for reporting deadlines. This will ensure that the general staff has the information necessary to ensure successful completion of the incident.

Determine the need for specialized resources. At incidents that are large and complex in nature, there is often a need for resources that are more specialized than standard fire and EMS equipment. At a mass casualty incident, resources needed may include buses, shelters, or even medical tents. The planning section chief must work with outside resources to determine the availability and any added value to the incident.

Establish special information collection activities. For the longer incidents, there may be a need to gather special information from outside sources. This information may include weather forecasts, research on hazardous materials, signs and symptoms of exposure to toxic substances, and other pertinent information. In the medical setting this information can assist in determining treatment or decontamination techniques, and prepare responders for potential issues with patients and responders.

Assemble information on alternative strategies. Because members of the planning section have access to information from the entire incident, the planning section chief holds the responsibility of developing contingency plans should the IAP need to be changed.

Provide periodic prediction on incident potential. The planning section chief is responsible for providing information to the incident commander on the progression of the incident. This may include potential problems or resource concerns as well as the expected success of operational objectives and tactics.

The planning section may also rely on technical specialists to fill the role at a larger incident. These individuals possess training that allows them to provide knowledge that a standard responder may not possess. In a mass casualty situation, a doctor may provide information on injuries or the need for a field hospital. In a flood, a public health official may provide information on potential issues associated with the flooding. Technical specialists are only activated when a need exists.

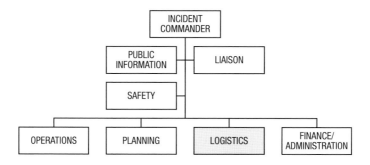

Logistics

The logistics section of the incident command system is responsible for all support activities associated with an incident. While recognition of the need for resources may come from the operations section chief, planning section chief, or the incident commander, the actual call for assistance and request of resources comes from the logistics section. Because of the range of resources that may be requested during an incident, the logistics section is further divided into two branches: service and support.

Logistics section chief. The leader of the logistics section is the logistics section chief. The individual chosen to staff the position of logistics section chief must have the ability to obtain resources and plan for their deployment and impact on the incident operations. He or she must also be able to organize and coordinate the fulfillment of multiple requests at once. The logistics section chief holds the following duties and responsibilities:

Assemble and brief branch directors and unit leaders. Once the logistics section has been staffed, the logistics section chief staffs the service and support branches, appointing appropriately trained branch directors. Once additional staffing is complete, the logistics section chief provides information briefings to the staff to ensure that they understand the operational objectives.

Participate in preparation of the incident action plan. The logistics section chief must participate in the development of the IAP and understand the operational objectives so that plans made can effectively support operations.

Identify service support requirements. The logistics section chief must be able to quickly identify the potential and actual needs of the operational personnel to ensure that the resources are on scene and available in a timely manner.

Provide input to and review communications, medical, and traffic plans. These plans, which are developed by members of the logistics section, must be reviewed prior to sharing them with the planning section and IC. This duty falls to the logistics section chief. The review includes determining whether or not the resources are available to support the plan and ensuring that it can be smoothly implemented.

Coordinate and process requests for additional resources. While a majority of the requests that will be received by the logistics section will be fire and EMS based, there is a potential for requests for items from other supporting agencies (e.g., towing companies, hazardous materials cleanup companies, etc.). The logistics section chief is responsible for the coordination and timely processing of all requests.

Review IAP and estimate section needs for next operational period. Along with ensuring that the IAP for the current operational period is current and the operational needs are met, the logistics section chief must plan ahead for the next operational period. This will ensure that requests for any potential future needs are started early.

Recommend release of unit resources in conformity with the demobilization plan. With the significant knowledge of the on-scene resources, the logistics section chief can make recommendation on which resources should be released. These recommendations are made based on the demobilization plan.

Service branch. The service branch is responsible for the objectives and resource requests that support the continued activity of the organization. Resource requests that fall under the responsibility of the service branch may include communications, food services, and medical care for on-site responders (not incident victims).

Communications unit. The communications unit is responsible for the appropriate and effective use of on- and off-site communications equipment and works to facilitate the procurement of appropriate equipment to support communications operations during incident response. Members of the communications unit are responsible for development and support of the incident communication plan and work to acquire and distribute equipment to support the plan. If developed, the communications unit also provides oversight to the incident communications center. The IC may choose to staff this position with an individual with significant background in incident communications and technologies. This may require requesting a subject matter expert from the locality's dispatch center.

Medical unit. The medical unit is an important function that an on-scene EMS provider may be asked to fill due to the nature of the work. As part of the service unit, the staff in the medical unit focus their attention on the responders. The medical unit develops the incident medical plan, provides first aid and basic medical treatment to incident responders, and develops procedures for major medical emergencies. These procedures may include triage, treatment, and transport of a large number of wounded responders (if, for example, a secondary event occurs). The medical unit is also responsible for the oversight of on-scene rehabilitation services. The responder rehabilitation manager position provides oversight to the rehabilitation unit. The responder rehabilitation manager is also responsible for development and implementation of on-scene rehabilitation services.

Food unit. The food unit is the third and final unit of the service branch. Members of the food unit are responsible for providing hydration and nourishment to incident personnel. In long-term incidents and incidents associated with severe temperature conditions, this is especially important.

EMS in the Logistics Section

While you may be used to focusing your medical skills on the treatment of victims of an incident, you must remember that you and your crew and other emergency responders to the scene have a need for medical assistance as well. The medical unit is responsible for the treatment and transport of on-scene personnel. This care includes the development and implementation of on-scene rehabilitation services.

The need to care for the providers on scene has become a high priority for all levels of incident response, from administrators to field personnel. Various studies of the risks of the job have found that nearly half of all firefighter fatalities and a major part of illness and injury are the result of stress and stress related issues. In order to lower these statistics and decrease deaths and injuries/illnesses, many organizations focus on educating emergency responders on health, nutrition, and physical fitness. They also work to standardize the delivery of on-scene rehabilitation services to ensure that responders have an opportunity to de-stress. One organization that has recognized the importance of on-scene rehab is the National Fire Protection Association (NFPA).

In *NFPA 1584: Standard on the Rehabilitation Process for Members During Emergency Operation and Training Exercises*, rehabilitation is defined as "the process of providing rest, rehydration, nourishment, and medical evaluation to members who are involved in extended or extreme incident operations."[14] The recognition is that rehab should be offered not only at emergency incident scenes but also at training evolutions, and that an emergency responder's body is put under the same stressors regardless of the purpose of the event. Because rehabilitation is viewed as a responsibility of on-scene EMS providers, it is important that you understand the regulations that support on-scene rehab as well as the benefits and potential activities of effective rehab. Written to support NFPA 1500, NFPA 1584 focuses on standardizing an organized approach to the delivery of rehabilitation services at emergency and nonemergency scenes.

Many agencies write very basic standard operating procedures (SOP) for rehabilitation operations. NFPA 1584 calls for relief from climatic conditions, rest and recovery, cooling or rewarming, rehydration, calorie and electrolyte replacement, medical monitoring, EMS treatment following local protocols, member accountability, and a process for release from rehab. These are relatively easy to translate into policy and procedure. The challenge and most frequent rehab stumbling block is span of control. Small and large departments alike often relegate rehab operations for large contingents of on-scene personnel to a single ambulance crew, a medical supervisor on scene, or a handful of cross-trained firefighters. Without sufficient rehab resources, the operation is certain to fail.

Two solutions have helped departments across the country implement effective and efficient rehab operations while maintaining the NIMS recommended 1:5 span of control. Remember, this span of control is a basic concept of ICS where ideally, each supervisor oversees five personnel (though the range can be from three to seven). One option to maintain appropriate span of control is to regionalize rehab operations, the other is to have firefighters rehab themselves. Both require careful consideration of available resources, considerable preplanning, and formal education of everyone involved.

Regional rehab typically uses dedicated teams who respond with a full complement of personnel, supplies, and equipment. With rehab and medical expertise, necessary equipment, and adequate staffing, regional teams have the ability to bring a fully functional rehab operation to any emergency scene and conduct rehab operations for the duration of the incident. Expenses can be shared among departments, borne by the host department, or funded through regional grants.

Allowing firefighters to assume ownership and responsibility for their own rehab is also a sensible approach to implementing a workable plan for rehab. NFPA 1584 calls for ongoing training of firefighters on fitness, hydration, nutrition, temperature regulation, and rest and recovery, all of which are important elements in the science of on-scene rehabilitation. This makes perfect sense given that firefighting is such a physically demanding and often dangerous activity.

Firefighters are truly performance athletes; to work at maximum capacity, reduce their risk of injury, and be as effective as possible on the job, they need to know as much about fitness and performance as any athlete. Such education should begin in the recruiting academy and continue throughout a firefighter's career.

With an understanding that properly conducted rehab would allow firefighters to spend more time on the fire ground and work harder and longer with less chance of injury, it seems logical that an educated, intelligent firefighter would want to participate in rehab. The duty of the department then, becomes providing a safe on-scene area, the necessary equipment and supplies, and available medical expertise to conduct rehab at incidents and training exercises. An EMS provider can observe firefighters entering rehab, assessing for any of the indicators specified in NFPA 1584 that suggest the need for immediate emergency medical treatment, allow firefighters to rest, cool or rewarm, and rehydrate with water and beverages provided, and then assess their own vital signs using a pulse oximeter or pulse co-oximeter (a device that also measures blood levels of carbon monoxide).

Firefighters who fail to rest and recover or who exhibit vital signs outside department parameters for return to the fire ground can consult with a medical provider in the rehab area for additional advice, assessment, and assistance. Those who respond appropriately to rehab and are ready to return to duty can check in with an EMS provider to be cleared from the rehab area. Such a process transfers ownership to the individuals who require rehab and benefit from it, reducing the ongoing interactions between EMS providers and firefighters in rehab to well within a reasonable span of control.

While the EMS provider designated to run rehab must be afforded considerable authority over decisions pertaining to release, implementation of rehab is the responsibility of the incident commander and decisions regarding rehab operation must be filtered through the appropriate chain of command. If you are part of a rehab operation, whether in a leadership or support role, you should ensure your activities support the operational objectives set forth in the IAP.

Support branch. The support branch is responsible for securing resources necessary to ensure the operations are uninterrupted. Resource requests that fall under the responsibility of the support branch may include facilities (port-a-potties, shelter, cots, etc.), supplies (medical equipment, etc.), and service equipment (figure 1–3). The support branch is also responsible for storage and inventory of all requested and received equipment.

Fig. 1–3 Trucks such as the one pictured can be used to provide rehab support during incidents.

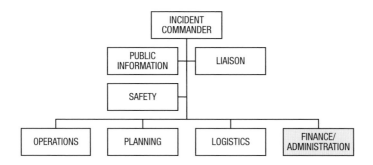

Finance/administration

The finance/administration section of the ICS structure is often left unstaffed during smaller incidents; however, during large/ complex/longer incidents, the section is an important aspect of incident response. The finance/administration section is responsible for accountability and the financial management aspects of incident response. This includes authorization of expenditures, as dictated by agency policies. It is important to note, though, that the actual procurement of any necessary resources is the responsibility of the logistics section.

The finance/administration section chief is the leader of the section. The individual staffing this position must have significant knowledge in the policies and procedures related to purchasing and administrative procedures. The section chief holds a number of responsibilities.

Manage all financial aspects of the incident. At activation, the finance/administration section chief takes control of all fiscal responsibilities of the incident, including spending and cost analysis. At smaller incidents, the section chief may choose to only activate certain functions of the section, such as cost analysis or cost reporting.

Provide financial and cost analysis information on request. At the request of senior agency and incident officials, the section chief must provide a cost analysis report. The cost analysis data can be used in estimates of the costs of future operational periods, as well as allow for the start of the reimbursement process.

Gather pertinent information from briefings with responsible agencies. The section chief must work with various agencies represented on scene to determine incident response costs and should be prepared to participate in operational planning meetings to determine additional fiscal and administrative needs of the incident.

Develop an operating plan and fill supply and support needs. Members of the finance/administration section are responsible for determining the activation status of resources during the incident. They also work with logistics section staff to ensure necessary resources are appropriately procured.

Meet with assisting and cooperating agency representatives. With the use of mutual aid assistance during incident response, it is imperative that the finance/administration section chief meet with mutual aid agency representatives to ensure that any costs and reimbursement issues are appropriately addressed and handled. Because of the ongoing work of the section chief, a deputy section chief may be assigned to serve as a liaison with the agencies.

Maintain daily contact with agency administrators on finance/administration matters. It is imperative that the finance/administration section chief continue to update the agency administration on the costs and expected costs associated with the incident. Reimbursement issues and funding may affect the resources that are activated or deactivated to assist with incident response.

Ensure that all personnel time records are completed and shared with home agencies. Time management and personnel timekeeping is an important part of the finance/administration section. In events that may eventually receive a federal declaration, timekeeping and appropriate records will assist in reimbursement of money spent. Time management and record keeping for individuals from outside agencies is overseen by the finance/administration section, but copies of those records should be shared with the home agency.

Provide financial input to demobilization planning. Many agencies may make decisions on which resources to demobilize first based not only on need, but also on the financial impact of the resources remaining on scene. The finance/administration section chief should work closely with the planning section to ensure that costs are provided to assist in the planning.

Ensure that all obligation paperwork is prepared and completed properly. Any commitments for reimbursement of funding that were made as a result of the incident must be met. Tracking and ensuring that these funds are paid is the responsibility of the finance/administration section. This includes ensuring that appropriate paperwork is completed and turned in to complete the purchasing process.

Brief agency administration on post-incident finance/administration issues. Completion of incident response does not necessarily mean that the role of the finance/administration section is complete. Because federal reimbursement can take time, the responsibilities of this section may continue for weeks or months after an incident is cleared. The section chief should continue to provide information to agency administrative personnel regarding expected post-incident concerns, including reimbursement and workers' compensation concerns.

Command Staff

A majority of incident personnel are assigned to a function within the structure of the general staff; however, three additional positions play vital roles. These positions are part of the command staff. Command staff personnel serve as assistants to the incident commander, directly reporting to that position. The need for staffing of the command staff functions is dictated by the size of the incident. Because these responsibilities fall on the incident commander when the positions are not staffed, it may be necessary to activate the command staff for a more effective and efficient operation. The command staff positions do not necessarily require medical knowledge, and as such can be staffed by other on-scene personnel. The individuals serving in the positions of command staff are known as officers.

Remember that regardless of whether a safety officer is assigned, every member of the command staff and every member of incident operations is responsible for ensuring safe operations.

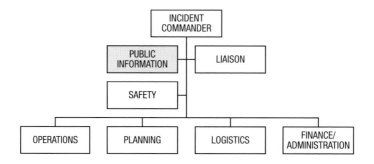

Public information officer

The public information officer (PIO) is responsible for effective and accurate dissemination of media information. An incident PIO may be the agency PIO or someone who has a good working knowledge of incident operations and a good working relationship with local media outlets. The PIO is the individual who coordinates the release of incident information through media briefings and press releases. The PIO does not release any information without approval from the IC.

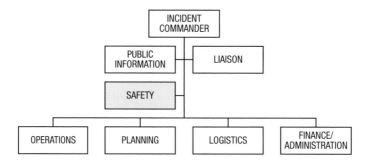

Safety officer

At any incident, whether it involves one ambulance or fifty, safety is the priority of all response personnel. Not only is life safety the primary concern of any incident, it is one of the first concepts taught in every EMS program, from basic to advanced. However, at an incident scene, when the need is determined to exist, the incident commander can assign an individual to serve as safety officer. The safety officer is a member of the command staff and, as a result, reports directly to the incident commander. The safety officer must monitor all phases of incident operations to provide information on witnessed safety risks and risk mitigation procedures. In relation to EMS, the safety officer may provide feedback to the incident commander regarding safe medical procedures, personal protective equipment (PPE) requirements, lifting and moving techniques, and even treatment of patients who have been decontaminated. The information provided by the safety officer may impact the incident action plan or medical plan written in support of incident operations. Remember that regardless of whether a safety officer is assigned, every member of the command staff is responsible for safe operations.

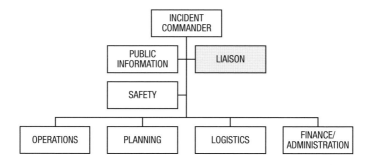

Liaison officer

Interagency cooperation is not always guaranteed when an incident requires the use of resources outside of the primary response agency. Prior to the implementation of the ICS concepts, poor communication between agencies, differing terminology, different operational goals, and a series of other limitations impacted the efficiency of mutual aid response in emergency incident operations. Even with the implementation of ICS concepts, there is still the potential of problems between different agencies, even from the same jurisdiction, working together. In order to facilitate the coordination of agencies at an incident scene, the incident commander may choose an individual to serve as the liaison officer. This individual serves as a conduit of information from the primary response agency and incident scene to outside agencies supporting response operations. These agencies may include other emergency response agencies (e.g., police, fire, EMS) as well as organizations such as hospitals, dispatch centers, and private resources (tow trucks, bus companies, medical supply companies). The liaison officer ensures that upon arrival on scene, outside agencies are provided with the appropriate point of contact.

TACTICAL WORKSHEETS

Each person holding a position in the ICS structure has expectations and duties he or she is responsible for accomplishing. It is not unimaginable to consider that a person could easily forget all of the job duties, especially while trying to function in an emergency event. To assist personnel assigned to ICS functions and ensure that all job duties and responsibilities are completed, tactical worksheets and checklists are available. Checklists provide a guide to ensure that the duties and responsibilities are complete, while tactical worksheets assist in tracking, recording, and disseminating information related to the incident. Tactical Worksheets are available from training.fema.gov/EMIWeb/IS/ICSResource/ICSResCntr_Forms.htm.

Implementation of ICS

Liability concerns

With the initial Presidential directive regarding implementation of a national incident management system, the focus was not on the effect of such a directive on liability of emergency responders. However, since its implementation and as a result of the rules and regulations put into place by many local and state governments, the liability of emergency responders has greatly changed. As a result of a ruling by New York's second highest court (in response to a case involving the death of a firefighter), fire, EMS, and police agencies are faced with an increased liability when NIMS standards are not put into place and adhered to during incident operations.

On March 7, 2002, the Manlius Fire Department in Onondaga County, New York, responded to a basement fire at a two-story residence. Upon arrival, an incident command structure was put into place and a county fire coordinator was called in to assist in resource management. After completion of vertical ventilation, Manlius fire personnel were reassigned to conduct fire suppression, as it appeared efforts to control the basement fire were not successful. After his arrival, the county fire coordinator took the time to assess the situation and upon recognition of uncontrolled fire conditions, testimony alleges that he ordered two firefighters to make entry into the structure. Based on the county fire coordinator's directive, two firefighters from Manlius Truck 2 entered the first floor while a third firefighter fed an attack line to them from the garage. Upon his attempt to make entry with the other two, the third firefighter found that the floor had collapsed and the other two firefighters from Truck 2 had fallen into the basement. Even with extensive rescue efforts of on-scene personnel, two firefighters perished.[15]

A lawsuit against the county fire coordinator and Onondaga County alleged that the firefighter's deaths were the result of the command given by the county fire coordinator. During the trial, lawyers discovered that once the county fire coordinator arrived on scene he neglected to communicate with the assigned incident commander. The belief was that because of his rank in his everyday position, the firefighters found it necessary to follow his orders even though during the current incident the county fire coordinator was NOT the incident commander. The plaintiff claimed that had the county fire coordinator not directed them to enter the building, they would not have died. The New York state court ruled that the failure of the fire department to follow NIMS served a basis for liability. The justification was that the NIMS documents "mandates a reasonably defined and precedentially developed standard of care, and does not require the fact's trier to 'second-guess [a firefighter's] split-second weighing of choices.'"

This ruling supports the theory that first responders and their agencies, whether career or volunteer department, may be held liable for noncompliance with NIMS requirements. The focus is on the use of the word "must." The word "must," which establishes a required mandate versus a voluntary mandate, is seen in multiple portions of the NIMS document. The concern is also that in creating documents to support the implementation of standard operating procedures (SOP) related to NIMS, local and state governments are unknowingly creating mandates that increase the liability of their first responders. Many recommendations have been made regarding implementation of NIMS to decrease liability concerns. They include creating guidelines instead of procedures that include an introduction explaining the purpose of the document and utilizing the word "should" instead of "shall" or "must." Along with some additional recommendations, these measures aim to ensure that readers and interpreters of the document recognize its specific purpose as a suggested practice, not a mandated practice.

Working toward compliance

With the concern about liability of whether your agency already has expectations of NIMS and ICS integration or they are just beginning to look at the use of NIMS/ICS procedures, the question always comes up regarding the best way to ensure that the use of ICS occurs in a seamless manner with little confusion. There are many actions that can be taken to help make the procedures and concepts associated with NIMS second nature to your providers. The first action is to use NIMS and ICS on a continued basis. Use of ICS in certain situations, for specific call types, may come natural to some. The suggestion, though, is that ICS be used in all situations, from a house fire or single patient medical call to a multi-agency response to a natural disaster incident. This allows for continued practice of the concepts, allowing personnel to become more familiar with the procedures. Another step in assisting in the smooth integration of NIMS into your response procedures is to provide training. NIMS training should complement current training practices. Training should be conducted frequently and in a variety of mediums including classroom, online, and exercise-based training. Training options are discussed in chapter 6. You should also review your state and local laws to determine whether any regulations impact the method of implementation or set specific requirements regarding the implementation of NIMS. Regulations can provide support for the implementation of NIMS programs and may also lead to the ability to support programs with financial support, a necessity for many departments when enacting policy changes.

Conclusion

With the needs of emergency services constantly changing, systems used to help with emergency response change too. From the original use of a command structure in California in order to better respond to wildfires, to the national standard of incident command you are familiar with today, incident management is a recognized need for efficient on-scene operations. Whether fire or EMS, career or volunteer, training on the incident command system structure is necessary for the most effective incident operations.

References

1. California Office of Emergency Services. "Some Highlights of the Evolution of the Incident Command System, As Developed by FIRESCOPE." March 26, 2003. Retrieved July 1, 2010, from http://www.firescope.org/ firescope-history/firescope-historical-documents.htm.

2. Chase, Richard A. *FIRESCOPE: A New Concept in Multiagency Fire Suppression Coordination.* General Technical Report PSW-40. U.S. Department of Agriculture Forest Service, May 1980.

3. National Wildfire Coordinating Group (NWCG). *History of ICS* (October 1994). Retrieved July 1, 2010, from http://www.nwcg.gov/pms/forms/ compan/icscomp.htm.

4. U.S. Department of Homeland Security, Federal Emergency Management Agency (FEMA) (April 2005). *NIMS Incident Command System for Emergency Medical Services, Instructor Manual.*

5. The White House, Office of the Press Secretary (February 28, 2003). "Homeland Security Presidential Directive/HSPD-5" [Press release]. Retrieved from http://www.fas.org/irp/offdocs/nspd/hspd-5.html.

6. The White House, Office of the Press Secretary (December 17, 2003). "Homeland Security Presidential Directive/HSPD-8" [Press release]. Retrieved from http://www.fas.org/irp/offdocs/nspd/hspd-8.html.

7. FEMA. "About the National Incident Management System (NIMS)," in NIMS Resource Center web site. Retrieved from http://www.fema.gov/ emergency/nims/AboutNIMS.shtm.

8. FEMA. "Ongoing management and maintenance," in NIMS Resource Center web site. Retrieved from http://www.fema.gov/emergency/nims/ OngoingMngmntMaint.shtm#item1.

9. Bourne, M. (December 1, 2005). "Need for NIMS." *Fire Chief.* Retrieved July 1, 2010, from http://firechief.com/preparedness/firefighting_need_ nims/.

10. U.S. Department of Homeland Security, Federal Emergency Management Agency (FEMA) (April 2005). *NIMS—Incident Command System for Emergency Medical Services, Instructor Manual.*

11. Ibid.

12. U.S. Department of Labor. "Hazardous waste operations and emergency response," in Occupational Safety and Health Administration (OSHA) web site. Retrieved March 9, 2011, from http://www.osha.gov/pls/oshaweb/owadisp.show_document?p_table=standards&p_id=9765.

13. FEMA (2008). *ICS-100. A: Introduction to ICS: Student Manual*, Version 2.0.

14. National Fire Protection Association. *NFPA 1584: Standard on the Rehabilitation Process for Members During Emergency Operation and Training Exercises*. 2008.

15. Pinsky, B. M. (March 1, 2009). "NIMS Directives and Liability." *Fire Engineering*. Retrieved July 1, 2010, from http://www.fireengineering.com/index/current-issue/fire-engineering/volume-162/issue-3.html.

FIELD USE OF ICS

Chapter Objectives

Upon completion of this chapter, readers will:

1. Understand the use and benefit of ICS in identifying leadership structure

2. Understand the use and benefit of ICS at a single patient/single unit incident

3. Understand the use and benefit of unified command

Scenario: "Welcome, rookie!" is the greeting you receive as you arrive to the station on your first day. You introduce yourself and shake hands with the paramedic in front of you, then place your gear by the ambulance and move inside. After putting your sleeping stuff on a bunk, you begin the round of introductions with the rest of the shift. From your orientation, you know the administrative command structure, but you weren't briefed on the operational staff organizational chart. Introducing yourself to the other shift members, you hear lots of new names and ranks.

"Thank God for name tags," you think. "I'm not sure I'll be able to remember everyone, much less who is in charge."

Considering the fact that there are five stations and three shifts, you realize there are a lot of personnel, including command staff, to remember. You do not know much about the organization of the operational staff. The one thing you do know is that you are the new guy, and are therefore the lowest on the totem pole.

How is anyone supposed to keep straight the operational chain of command, especially in a large organization? What can be done to ensure that you, as the new hire, and others coming in after you, can keep the leadership and organizational structure straight?

Introduction

It does not matter whether you are a probationary EMT working your first day on the ambulance or a rookie reporting to the engine for the first day at the station, you need to know who is in charge. This provides guidance in the station house as well as guidance in the field. For both career and volunteer staff, it is important everyone is aware of the chain of command and can work within the structure during emergency response. In the development of the national incident management system and the incident command structure, there was recognition of the need to integrate the concepts not only at emergency scenes, but also on a day-to-day basis. This assists in a smooth transition for personnel from the everyday activities of an emergency response agency to the on-scene operations at an emergency incident.

Daily Use of ICS

Daily use of the incident command structure (ICS) is the best way to ensure that every provider is familiar with the concepts of ICS, including chain of command and span of control. Consensus among fire personnel is that the more ICS is used, the more comfortable response personnel become with its use in emergency situations. The question becomes, how do you integrate a system designed to organize response at emergency scenes into daily activities? The answer is "carefully."

Using ICS to identify leadership

Most agencies, yours may be included, already utilize common terminology to designate leadership positions. The terminology is the same across the board, regardless of the station you are assigned to. Master firefighter, sergeant, lieutenant, captain, engineer, and chief (battalion, EMS, etc.) are terms used to provide a rank structure within an organization. Based on many military structures, this provides not only leadership identification, but also a career path for personnel to achieve. The problem becomes that leadership positions in the everyday use are not regulated. For instance, department A may have the rank structure shown in figure 2–1. Department B, which is in the next county over, may have a rank structure similar to that in figure 2–2. When these departments work together in a mutual aid incident, leadership conflicts may arise due to a disconnect between each department's terminology and leadership structure.

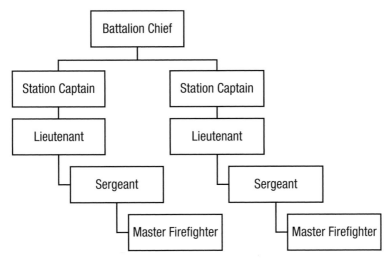

Fig. 2–1 Department A sample ranking structure

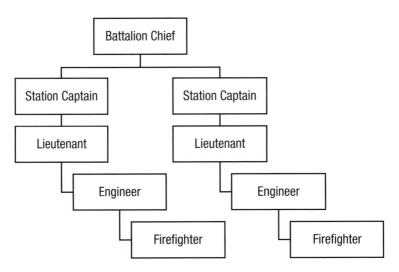

Fig. 2–2 Department B sample ranking structure

Many departments look at various positions to predetermine who will staff the ICS positions during an emergency event. Potential conflicts arise when position roles are labeled differently between jurisdictions, especially when determining who has higher authority in staffing positions. Understanding each department's rank structure can assist in understanding which personnel are the best option for staffing various positions within the ICS structure. Departments that work together through mutual aid, most especially those that border each other, should ensure that leadership structure is similar, if not the same, in terminology to ensure that implementation of the ICS structure and filling of ICS positions is smooth during an emergency event. Remember, though, that in the long run you should consider training and experience as the primary factors in determining the appropriate personnel for ICS positions.

Scenario: "Good morning, guys," you say to the outgoing shift members. "How was the night?"

It has been a few months since you received your assignment for this station, and while you are beginning to develop a routine, there are still some things that seem a bit off. Being the new member of the shift, you keep your mouth shut and imagine that it will continue to get easier. Every shift begins with a general discussion with the off-shifting personnel. It gives you a chance to hear about the call load from the previous shift, any potential equipment and apparatus problems, and other information they may have come up during the shift. After the off-shifting personnel leave, your shift sits around drinking coffee and catching up on what each person did on his or her day off. From there it is normally just a matter of waiting until the first call is received. Today something a bit different occurs.

"Shift meeting in 5 minutes in the day room," your supervisor says over the station intercom.

"Great! What did you do now?" members of the shift joke, looking in your general direction.

During the meeting, the supervisor begins to relate some changes that have come down from administration. Most of the changes include a greater integration of ICS principles within the daily routine of department personnel.

What changes do administrative personnel believe they can make to assist in the daily activities of the agency and operations personnel? Is there truly a benefit to using ICS on a daily basis, when its application is a bit different in the field? Will this really provide any order, or will it cause more confusion regarding daily operations and command structure?

Daily use of terminology and concepts

One of the first steps in ensuring that ICS smoothly transitions into your daily activities is using the appropriate terminology. As discussed in chapter 1, one of the benefits of ICS is common terminology. Common terminology should be stressed among people, stations, and even neighboring jurisdictions. In many large urban areas, departments come together to create a bank of common terms. For departments that have adopted national incident management system (NIMS) principles and are working toward NIMS compliance, the federal government offers a list of standard terminology relating to equipment and personnel. Through resource typing, the Federal Emergency Management Agency (FEMA) has created a minimum standard for resources.[1] While the original intent was to ensure that resource requests during a federal response met a specific standard, resource typing can assist departments in both terminology of apparatus as well as ensuring that an apparatus meets recommended capabilities. Along with typing of EMS resources, FEMA has also typed the following categories of resources:[2]

- Animal health emergency

- Fire and hazardous materials

- Incident management (IM)

- Law enforcement

- Mass care

- Medical and public health

- Pathfinder task forces

- Public works (PW)

- Search and rescue (SAR)

While your work in the EMS field does not require the use of some of these resources, understanding their capabilities may assist in some aspects of your job in the field.

The following is a list of EMS resources that have been typed by FEMA:

- Air ambulance (fixed-wing)

- Air ambulance (rotary-wing)

- Ambulances (ground)

- Ambulance strike team

- Ambulance task force

- Emergency medical task force

- Mass casualty support vehicle

- Multi-patient transport vehicle

Chain of command and span of control. Two more principles that can be utilized during daily operations that are also emphasized in the ICS structure are well-defined chain of command and appropriate span of control. Most departments have a well-defined chain of command they incorporate in their daily operations. The department chief or director serves as the "commander." As with the incident commander at an incident scene, this individual holds ultimate responsibility for decisions made within the department. Each position has a supervisor, and potentially serves as the supervisor to others. Each position also understands to whom they report and how information can flow. By enforcing chain of command principles on a daily basis, personnel are more likely to understand and follow these principles at an incident scene. The chain of command within an organization is something that should be taught to personnel during an orientation process to ensure that they know it coming in. While names may not remain the same, the position titles and responsibilities of each position should be reviewed.

Chain of command is often utilized in the most simple of assignments on a daily basis. For example, many departments use chain of command when making riding assignments for the apparatus (figure 2–3).

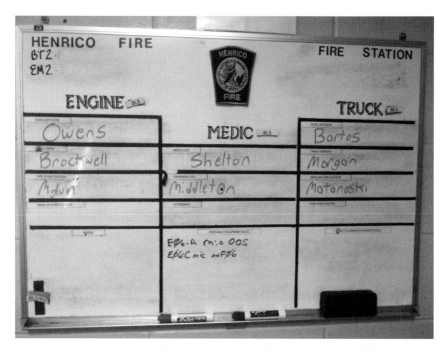

Fig. 2–3 Making riding assignments at the beginning of the shift ensures that personnel know who is in charge and what role they fill during the shift.

The fire apparatus normally has an assigned officer (or acting officer if an officer is not on shift), a driver, and a firefighter. The officer is the unit leader, providing supervision to the other personnel. In an ambulance crew, one person serves in the role as attendant-in-charge. This is the individual serving as the unit leader. At the beginning of each shift, with unit assignments made, the unit leader can set forth expectations and begin the process of accountability for his direct reports.

Span of control should also be maintained on a daily basis. According to ICS principles, a position should supervise from three to no more than seven people at a time, with a recommendation of one supervisor for five reports. Most organizations ensure this occurs through the use of various levels of authority. Sergeants report to lieutenants, who report to captains, who report to battalion chiefs. Although the terminology of these positions may vary from department to department, the understanding of who your supervisor is and what position he or she holds is the important part. If span of control does begin to fall outside the appropriate range, departments look toward reassignment of personnel or appointment of additional supervision. In other words, just as would occur at an incident scene, the organizational chart expands.

Management by objectives. At an operations scene, management by objectives is easy to accomplish. The goals are well established and training of personnel often focuses on how to meet the objectives at a standard emergency incident. Management by objectives is a bit harder to conceptualize in the everyday EMS environment. Consider the scenario at the start of this section. Aside from the initial discussions with off-shifting personnel, there appears to be no structure in how the shift runs. A supervisor who manages by objectives may provide on-shift personnel with daily expectations, a "to-do" list of sorts. This may include completion of daily check sheets, participation in community outreach (e.g., business visits, preplanning, etc.), or training exercises. By creating a list of objectives, a supervisor is able to determine which objectives are completed and which are not. He or she may also be able to analyze the shift's activities and determine the reason behind tasks not being complete. This enhances the effectiveness of the functions of the organization. As with an emergency scene, if a supervisor determines that objectives are not being met, it is his or her responsibility to determine why, and make any necessary changes to either the expectations or the personnel.

Unit integrity. In many departments, unit integrity is key to daily operational activities. As discussed in chapter 1, unit integrity is the ability to maintain crews within their group and assign them to functions based on their knowledge, skills, abilities, and equipment. In the fire department, unit integrity begins the minute station assignments are made during rookie school. Most new firefighters are assigned to the engine company. Additional assignments include the truck company or the rescue squad/company. These assignments allow personnel to understand their roles at an incident scene. The engine company focuses on fire suppression, the truck company on rescue and ventilation operations, and the rescue squad/company focuses on specialized operations (i.e., water, hazardous materials, vehicle, and other special incidents). In the EMS field, the assignment may be as simple as a basic life support (BLS) or advanced life support (ALS) truck. These assignments focus on the training level of the provider. Often, a BLS and ALS provider are paired together, making the truck capable of handling basic and advanced level calls. While unit integrity may not be based on an assignment to a specific company, when providers are assigned to the ambulance they understand that their main function is to provide patient care at the level to which they are trained.

The everyday use of the ICS system and integration of its characteristics within daily operations provides for a great opportunity to become familiar with it for use in emergencies. It is also important to remember that ICS, while highly beneficial at large-scale incidents, is also beneficial at smaller, simpler incidents.

Scenario: It has been a good shift. Fairly quiet, with just enough action to keep things busy, but to give you a few minutes of rest in between as well. You and your fellow crew members have just completed dinner when the tones drop.

"Medic 1, respond to 1497 Johnson Street for a general illness call."

As you put the last dish in the dishwasher you grab your radio mic. "Dispatch from Medic 1, we are enroute."

"Medic 1, be advised you are responding for a 69-year-old male complaining of vomiting, general weakness, and diarrhea. Patient states he has been vomiting on and off all day."

"Medic 1, copy," you radio as you refer to the GPS monitor and confirm the location of the call. You turn to tell your partner what you need upon arrival at the scene.

"I thought I was running this one. You ran the last call," your partner states to you as you pull into the driveway of the residence.

What can be done to prevent confusion on a call? Is utilization of the ICS structure necessary at a call with only one patient and one piece of emergency apparatus? If so, how can ICS help ensure that the call is organized and that patient care is provided in the safest and most efficient manner?

Use of ICS in a Single Patient/ Single Unit Response

In EMS, the single patient, single unit calls are the most common. You arrive on scene with your partner and provide care from start to finish, without assistance from other crews. You handle the patient without hesitation and manage your duties through the working relationship you have developed with your partner. What happens when that relationship falters and organization of the simple, single patient call begins to fall apart? It may seem like a bit of overkill, but the use of ICS at a single patient incident can provide guidance and practice in utilization of ICS principles and practices (figure 2–4).

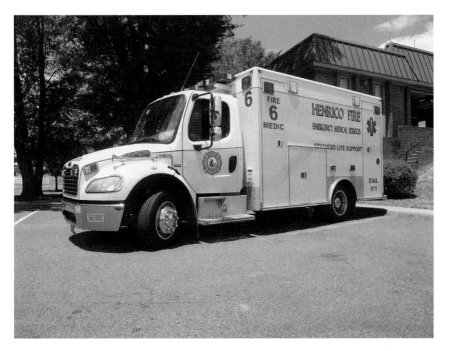

Fig. 2–4 While uncommon, developing an ICS structure on a single patient/single unit response can assist personnel in knowing their roles.

When apparatus assignments are made at the beginning of the shift, one person is assigned as attendant-in-charge (AIC). In a truck with both a BLS and ALS provider, the role may change back and forth depending on the level of treatment required for the patient. This is something that must be decided at the start of the shift so there is no confusion who is running which calls. With the knowledge of who will serve as AIC, you have also determined who will serve as the incident commander during the calls. As the first arriving unit, you should conduct yourself the same, regardless of incident size, by providing a scene size-up.

Scene size-up

Your scene size-up is the opportunity to provide a visual description of the scene to other responding units or to the dispatcher who may have to send additional units to assist you. It provides an overview of incident disposition. A scene size-up should allow personnel not on the scene to understand the basics of the incident scene. Effective scene size-up is stressed to firefighters in training because it is most often utilized when arriving at the scene of an alarm or working fire. In the EMS field, conducting scene size-ups and communicating pertinent information over the radio can be useful as well. While the components of a scene size-up are often taught in basic EMS programs, the use of them in radio communication is not normally stressed. As a step in the process of integrating ICS as standard practice for every call, effective scene size-ups need to be communicated over the radio.

A good scene size-up consists of five components that, when completed, can assist in initial scene control and recognition of the need for additional resources. The first component is determining the number of patients.[3] You must immediately recognize the number of patients involved in the incident as this can be a major factor in determining the need for additional resources. An incident that results in more patients than there are providers to treat them is deemed a mass casualty (multi-casualty) incident. These types of incidents will be discussed in more detail in chapter 3. Once you recognize the number of patients, you must determine the mechanism of injury (MOI) or nature of illness (NOI), the second component. The MOI/NOI may be gathered from the dispatch information. In the scenario at the start of this section, you may determine from the dispatch information that your patient is suffering from a viral or bacterial infection, making it a medical call and meaning a medical assessment is necessary. In a trauma call, the MOI can provide vital information on the potential injuries your patient may experience and also designates the need to complete a trauma assessment. Determination of the MOI/NOI can also assist in determining whether or not you can provide the level of patient care required for the situation.

The third component of an effective scene size-up is determining whether or not you have the appropriate resources. The need for additional resources may have already presented itself as a result of fulfilling the first two components. For instance, if your patient is suffering from chest pains or significant multi-systems trauma and your truck is a BLS truck, then an ALS provider is needed. Regardless, recognizing the need for additional resources quickly and making the request can assist in the efficiency of your call. The fourth component of a good scene size-up is determining the standard precautions or necessary personal protective equipment (PPE) required.[4] PPE needs may be as simple as a pair of gloves or as complex as a need for dispatch of hazardous materials response teams and patient decontamination equipment. Regardless, the scene size-up provides an opportunity to make this recognition and take steps to ensure that PPE levels are appropriate before beginning patient care.

The last step of the scene size-up process is determining whether or not the scene is truly safe. Each of these steps is ongoing, as the need for resources, the appropriate level of PPE, or the safety of the scene has the possibility to change at any time throughout the incident. If you cannot ensure the safety of yourself (primary), your crew (secondary), and bystanders and the patient (tertiary), then you should not be in the scene. Making a scene safe may involve some of the other components of scene size-up. Scene safety may be achieved by calling additional resources (based on the recognized problem: fire, police, animal control, etc.) and by selecting and donning the appropriate PPE. If a scene is not safe or becomes unsafe, quickly remove yourself from the situation until the scene is safe again. Remember that scene size-up is a continual process. Because the incident may change (i.e., become unsafe, increase in number of patients, require additional resources), you should continually monitor the scene and determine any change in your needs. Also, remember that once your initial scene size-up is completed, pertinent information should be communicated over the radio.

Establishing the command structure

With a scene size-up completed, the attendant-in-charge (AIC) must establish the incident command structure. As assigned AIC, this task may be as simple as reminding your partner of his or her responsibilities during the call. Remember, as discussed in chapter 1, when a role is not filled by a person, it becomes the responsibility of the incident commander to ensure that the job duties are carried out. In a single patient/single unit response, you, as the AIC, may serve as both the incident commander and the operations section chief. As the AIC you are in charge of patient care and therefore are making the operational decisions, as well as ensuring that everything else is being completed. Your partner fills the role of logistics section chief. He or she is responsible for getting equipment, ensuring that you have the resources you need, and requesting additional resources when directed, much like the role of a logistics section chief. Because there is not an expectation for planning or financial services, those are roles that do not need to be filled. If the incident scene changes, those roles should be filled. Figure 2–5 shows a sample command chart for a single patient/single unit incident.

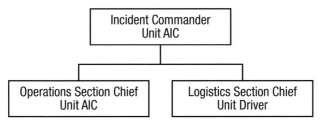

Fig. 2–5 Sample command chart—single patient/single unit incident

Using the scenario at the beginning of this section, radio traffic for a scene size-up may be as follows:

"Dispatch from Medic 1, we are on-scene of a single story structure with one patient. Medic 1 officer will have command."

Scenario: You are startled awake by the sound of the tones and the lights coming on. The dispatcher's voice comes across the intercom.

"Medic 4-2, Engine 4, Squad 2, you are responding to the intersection of River Ferry and Woolwine Road for a single vehicle accident."

After wiping the sleep out your eyes, you and your coworkers head down to the apparatus bay to respond. After marking en route, the dispatcher comes back with additional information.

"All units, you are responding for a single car motor vehicle collision at this intersection. The caller states that the driver is still in the car."

You realize that you will probably beat Engine 4, since they were actually coming back from an earlier alarm call. As you pull up to the scene you realize you are the first on scene.

"Dispatch from Medic 4-2, we are on scene with a single vehicle collision, vehicle on its side. Be advised that the occupant appears to be trapped in the vehicle. We will need additional law enforcement to control traffic. Medic 4-2 officer will take incident command."

As you finish your radio traffic, you see both Engine 4 and Squad 2 arriving on scene.

With additional units arriving on scene, what is your first priority? What impact does assuming the role of incident commander have on your ability to provide patient care? Are there alternatives to taking command, knowing someone better suited for the position will arrive shortly? Knowing that this is more complex than a single patient/single unit call, what does the ICS organizational chart look like?

Use of ICS in a Single Patient/ Multiagency Response

The above scenario represents events you are faced with repeatedly. While you may not always be the first on scene, you are often faced with the need to work with additional companies during the response to an emergency. It is normally just one or two other units, but it still provides an opportunity for interaction with other providers and use of the incident command structure to maintain order and ensure that the patient receives the best care (figure 2–6).

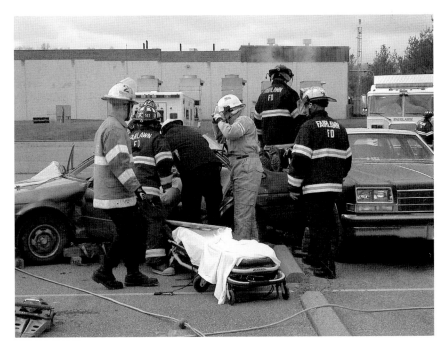

Fig. 2–6 A car crash requires the response of fire, police, and EMS resources. The ICS structure for a multiagency response will be different from the structure for a single unit incident response.

Passing command

For any call, the first arriving unit should provide a good scene size-up and the officer or a member of the crew must determine whether it is appropriate to take command. If you feel that it is not appropriate and a more suitable person to fill the role of IC is en route, then you can pass command.[5] Passing command should occur only when there is a compelling reason. In the above scenario, passing command could occur due to the need for the medic officer to focus on patient care. The decision to pass command should be communicated in your initial radio traffic. For instance, in the above scenario, Medic 4-2 could have provided the following communication:

> "Dispatch from Medic 4-2, we are on scene with a single vehicle collision, vehicle on its side. Be advised that the occupant appears to be trapped in the vehicle. We will need additional law enforcement to control traffic. Medic 4-2 is passing command to Engine 4."

With that radio traffic, the officer on Engine 4, which is the next arriving unit, is made aware of the fact that he will need assume the role of incident commander on arrival. Engine 4's officer must also acknowledge this over the radio. Upon arrival, Engine 4's officer will then assume command.

Establishing and transferring command

As with passing command, the decision to assume command must be communicated during your initial scene size-up and concurrent radio traffic. While you may not necessarily maintain the role, it is an important step in quickly gaining control of the situation. In the above scenario, because there are multiple actions that need to be taken (i.e., patient care, extrication, hazard control), it is reasonable to believe that the role of incident commander should be transferred to a member of Engine 4's crew. This allows you and your partner to focus on patient care instead of scene management.

Remember that many states have requirements regarding who serves as the command at an emergency; in not following those requirements you are facing significant liability.

Even if the scenario described an incident where the fire department was dispatched to support patient care (high priority medical call), incident command is still best held by a member of the engine crew. Remember that the focus of the EMS provider should be patient care. However, if you must take command, you should work to ensure that patient care is handled by other arriving EMS providers. Establishing command is a simple task. Once you have arrived on scene you should conduct a scene size-up and provide that information over the radio. By assuming command you take on the initial responsibilities of the incident commander. If additional units have not yet arrived on scene, then you need to consider conducting the initial activities of the incident commander. First, determine if there is a need for a physical command post. If so, provide that information to incoming units so they know where to go to upon their arrival. You should also determine the need for additional resources. If that appears to be a necessity, then make the request. Once an individual who is more qualified to serve as the incident commander arrives on scene, you need to be prepared to transfer command.

Transferring command may occur for four reasons.[5] The first, as discussed, is that a more qualified person arrives on scene to assume command. The second is that a legal requirement forces the change. The third is due to staff changes during a long-term incident. Because shift changes can occur during incidents of long duration, the incident commander position will need to be transferred from off-going shift members to oncoming responders. The last reason for transfer of command is more prevalent in larger incidents such as a disaster or mass casualty response. This reason for transferring command is that the scale of the incident has decreased and the home agency is now capable of finishing response activities. In this case, transfer of command is being done because resources have been released and the home agency is once again the primary, if not only, response agency involved.

Once you have decided to transfer command you must go through the appropriate procedures to ensure that operational aspects of the incident are not negatively impacted while command is being transferred. The first and a most important aspect of command transfer is the command briefing. In order to ensure that the incoming incident commander has full knowledge of the incident, you must provide him or her with a complete operational briefing prior to exiting the incident scene or being reassigned. The command briefing should be conducted face-to-face to ensure that there is no miscommunication. Your command briefing should include the following information:

- Tactical priorities

- Action plans

- Hazardous or potentially hazardous situations

- Accomplishments

- Assessment of effectiveness of operations

- Current status of resources

- Additional resource equipment

If incident operations have not been underway for a long period of time and few activities have been undertaken, the command briefing may be short. Once a command briefing has been conducted, the transfer of command should be announced over the radio. Radio traffic should include the end of the command by you and announcement of the name of the new command officer. With a new IC established it becomes imperative to provide communication to on-scene and incoming personnel. The new IC should establish and share tactical priorities, action plans, hazardous conditions, status and effectiveness of resources, and determine a need for additional resources. This provides an incident update to all personnel and also communicates any changes in the priorities or actions as a result of the new IC.

Establishing the command structure

When all units have arrived on scene at a single patient/ multiunit incident, a command structure can be fully established. Whether transferred or passed to the incoming engine company or command officer, command is not your responsibility. As the only EMS resource on scene, your responsibility is patient care and any other aspect of emergency medical care. Because there are few people on scene, the ICS organizational chart may be designed to focus on the specific operational activities. Figure 2–7 provides an example of an organizational chart that may be utilized to organize response efforts.

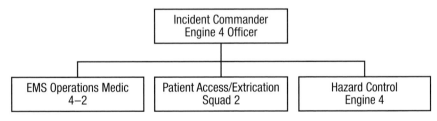

Fig. 2–7 Sample ICS chart for response to single patient/multiunit car collision

While the scenario in this section is a call requiring fire and rescue operations, the possibility exists that multiple units may respond to an incident that only involves patient care and movement. For instance, consider the following scenario:

Scenario: You are staffing a BLS ambulance. You and your partner are dispatched to a cardiac arrest at a residence. You respond, along with Engine 1 and EMS-1 (the on-shift ALS provider).

In this scenario, the fire crews will be directly involved in patient care and movement as there are no rescue or firefighting needs. Because you still need to focus on care of the patient, the engine officer should still hold the position of incident commander. Figure 2–8 provides a sample organizational chart for this scenario. Because EMS-1 is the higher trained EMS provider, he/she should serve the role of EMS group supervisor, providing direction and expectations to the other support personnel.

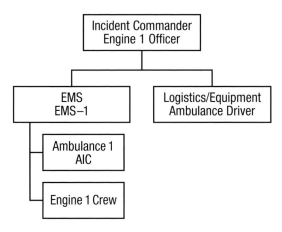

Fig. 2–8 Sample ICS chart for response to single patient/multiunit cardiac arrest

When multiple patients exist, there are obviously more EMS personnel to integrate into the organizational chart. Remember that one of the benefits of using ICS is that the organizational structure can expand and contract to meet the needs of the incident. If the incident involves more than one ambulance, then they are appropriately placed under the supervision of the EMS group leader (see figure 2–5). This maintains the basic characteristics of ICS such as unit integrity and span of control. As a reminder, when the patients begin to outnumber providers, new positions within ICS become important in the organizational and response process. Chapter 3 provides more detail on this type of situation.

Scenario: As the EMS supervisor, you have spent much of your time in the quick response vehicle (QRV). You are looking forward to making it back to the station for lunch when your tones are set off.

"Engine 2, Engine 3, Squad 1, Medic 3, Medic 4, EMS-1, Battalion 2, respond to the intersection of St. James and Allen Streets for a school bus collision. Law enforcement on scene advises that a school bus has collided with a light pole at this intersection."

Prior to your arrival Battalion 1 has arrived on scene and assumed the role of incident command.

"All units, be advised that the command post will be at the back of my vehicle parked just west of the intersection," Battalion 1 states.

Upon your arrival, you approach the command post and request an assignment. As you look around, you notice that the police officers have gathered to the east of the intersection around a command vehicle as well. Meanwhile, traffic is still trying to get through the intersection, which is still partially blocked by the bus.

"EMS-1, I need you to establish an EMS sector and begin managing the patients. We need to get a handle on how many students are actually in need of medical care," the incident commander says to you. He continues, "Engine 2 officer, I need your crew to begin putting out flares, establishing a scene perimeter, and start traffic control. Once this hits the news I can only imagine how many parents are going to try to come directly to the scene."

While you can understand the need for traffic control, you cannot help but wonder if this is going to be a duplication of effort, considering that is normally law enforcement's responsibility.

What can be done when multiple agencies and jurisdictions are involved in emergency response to ensure that they work together to meet the needs of the incident? How do you ensure that agencies talk to each other as they work at the incident?

Unified Command

You may find yourself in a situation where not only have multiple companies within your jurisdiction responded to a call, but response agencies from other jurisdictions are assisting in the response as well. When this type of situation arises, other command structures are best suited for organizing this type of incident response. Each of the agencies responding and each of the jurisdictions assisting in the response have a desire to provide input in strategic and tactical decision-making. For these types of situations, organization of responders using the unified command structure is best.

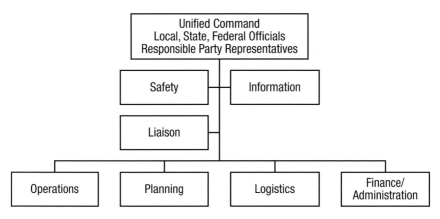

Fig. 2–9 Unified command structure

Deciding to implement unified command

As you may recall from chapter 1, the unified command (UC) structure is utilized when multiple jurisdictions and agencies have a voice in the decision-making process. Representatives from each agency gather at the incident command post and provide input into the objectives and strategies of the incident to ensure that their responsibilities are addressed.[6] While the agency representatives are all involved in the decision-making process, it is important to remember that the ultimate authority falls on the incident commander, and there is still only one person in that position. The individuals serving as agency representatives should ensure that they have decision-making authority, as they will be speaking as the voice for their agency. This will ensure that decisions occur as quickly and efficiently as possible.

When you arrive first on the scene of an emergency incident, it becomes your responsibility to begin the development of a command structure. The previous incidents (single patient/single unit, single patient/multiunit, multipatient/multiunit) have all provided a great opportunity for utilization of the standard IC structure. The single IC adequately represents the on-scene resources. For some incidents, however, it becomes immediately apparent that a unified command structure is the most beneficial organization for resource and response management. The first, as previously stated, is an incident that affects more than one jurisdiction.[7] This could be as simple as a highway incident that occurs at the county line or as complex as a pandemic outbreak within multiple jurisdictions. A second incident that should be immediately recognized as benefiting from a UC is an incident that affects multiple agencies within a single jurisdiction. An example of this is a hazardous materials incident. In this response, fire is responsible for fire and hazardous materials control, police is in charge of evacuation and scene control, EMS is responsible for patient care, and a hazardous materials cleanup company is involved in cleanup activities. A third incident involving the application of a UC is an incident that crosses multiple jurisdictions and affects multiple functional agencies. Incidents like this can involve both natural and man-made events (e.g., terrorism, hurricanes, floods, tornados, etc.). Response to these types of incidents involves local, state, and federal agencies through a multi-jurisdiction area. The last type of incident benefiting from a UC is an incident that affects multiple levels of government (local, state, and federal). An example of this type of incident is a pandemic event. This involves local, state, and federal health and human services officials (including EMS agencies) working closely together to respond to the event. In this type of incident, all levels of government must be involved in the decision-making and response process, making a unified command structure imperative.

BENEFITS OF THE UNIFIED COMMAND ORGANIZATION

The decision to develop a unified command organization instead of a single command structure may seem simple when you look at the type of incident, but in terms of actual structure, its advantages may not be as clear. It seems counterproductive to include more people in the decision-making process. Consider how long it takes your crew to determine where you want to go to dinner each night, and that does not involve the need to make a split-second decision. There are many benefits to a unified command organization that you should consider when determine which organizational structure to use.

The first benefit is that a single set of objectives guides the responders at the incident. If each responding agency developed its own command structure, each would develop a set of objectives to manage their response. These objectives may conflict with other responding agencies, causing costly delays in response efforts.

Another benefit of a unified command structure is that with a coordinated development of strategies and objectives, duplication of response activities decreases. When agencies work together to determine response activities they can assign responsibility to the appropriate agency, without assuming that it is theirs. A third benefit of a UC structure is that the flow of communication is the same, regardless of which agency is providing the information. This ensures that each agency receives and disseminates the same information, decreasing the potential of operational personnel receiving conflicting reports.

Developing an effective unified command structure also ensures that the authority and power of each represented agency and jurisdiction is maintained throughout response activities. With agency representatives involved in the development of objectives and strategies, individual authority is maintained.

While there are many benefits to using a unified command structure, it is important to recognize the need for the structure early and begin implementation immediately. This will ensure that the minimum time is spent between the beginning of response activities and arrival of appropriate agency representation in the incident command post.

Guidelines for implementing unified command

While the incident type has a huge impact on whether to implement a unified command or single command structure, there are also some guidelines to keep in mind for the unified command structure to be successful. The first guideline is familiarity with the functions of ICS unified command. While emergency responders are heavily trained on standard ICS functions and application, there is less focus on the concepts associated with unified command. With this in mind, implementation of a unified command structure requires that personnel be intimately familiar with the concepts, characteristics, and features to ensure a smooth implementation. A jurisdiction is also more likely to make the decision to participate in the unified command structure when they are familiar with the benefits of the system. Another key to the implementation of the unified command structure is that facilities that support response are collocated. As previously stated, not only does collocation help prevent duplication of activities, but it also allows for a coordinated effort and efficient deployment of resources.

Another feature of unified command is each member of the system must agree on who is responsible for implementing portions of the incident action plan. While every agency's objectives and priorities are kept in mind, agreement must be reached regarding which agencies are responsible for which priorities. Multiple agencies may have the resources to complete specific tasks, but which agency is *most* responsible for completion of the task is important to consider. It is also important to designate one individual to serve as the public information officer (PIO) and one agency representative to serve as the point of contract for the operations personnel. The PIO speaks for all agencies represented at the incident, normally relying on press releases developed by a collective group. Selecting an agency representative to serve as a single point of contact for the operations personnel ensures that all information is coming into the unified command group and that outgoing information to each section is consistent. The last feature that ensures a smooth implementation of the unified command system is that individuals that will serve in the unified command structure should train together, ensuring that what they practice in a training exercise is appropriate so that once implemented, it is done smoothly and with little confusion. Training is discussed in chapter 6.

Elements of unified command

With the determination of the need for a UC structure, the four elements of unified command must be implemented. The first element is policies, objectives, and strategies.[6] These are preexisting policies that departments and agencies rely on to provide input during the incident. Similar policies, objectives, and strategies may exist between various response agencies, but each agency's representative will make decisions according these items. The second element of UC is organization. Organization is provided to the UC structure as long as the agency representatives fit within the predefined system. The recognition of their positions and ensuring that they follow the principles of ICS (chain of command, unit integrity, etc.) will ensure that organization is maintained throughout the incident. The third element of unified command is the resources. Each representative agency within the incident command post must have resources that they provide to the incident. An agency that does not provide resources to incident response does not need to sit in the

ICP. The fourth and final element of the UC structure is operations. With the policies, procedures, objectives, and resources all determined, an operations section must exist to put into tactical processes the decisions made by the representatives in the unified command structure. With the four elements of the unified command organization met, decision-making and strategy development becomes a more effective and efficient process.

Unified command features

As with a single command structure, organization using a unified command structure presents with certain features. These features are what support the organization and its functions and ensure that the elements of the unified command structure are maintained. The first feature is that there is a single incident organization.[7] As previously stated, this method of incident command can lead to duplication of response efforts and resource requests. When a UC structure is maintained, the appropriate individuals serve together in the incident commander position.

A second feature of a unified command organization is the use of shared facilities. By ensuring that facilities are shared by each response agency, you are ensuring that efforts are coordinated at all levels of organization. The shared facilities begin at the incident command post, with all representatives serving in one location, and continues on through the entire ICS structure, including the staging area and other on-scene facilities. A third feature of the UC structure is that there is a single planning process and incident action plan (IAP). This feature ensures that each response agency and jurisdiction works from the same plan and toward the same goals. This means that there is no freelancing or competitive activity occurring during incident response. Each agency works together to achieve the same outcome.

A command meeting is necessary to ensure that the IAP incorporates the objectives of each agency and that representatives can communicate with their responders the priorities and tactical assignments to ensure a unified response effort.

THE COMMAND MEETING

The command meeting is held before the first operational period, further strengthening the need to develop the UC structure quickly. The meeting provides an opportunity for all agency representatives (agency IC) to provide input and participate in the development of the incident action plan (IAP). The information provided and discussed during the command meeting is utilized during the IAP meetings before each operational period.

The command meeting is an opportunity to gather information on agency/jurisdictional priorities and objectives and limitations, concerns, and agency restrictions. This assists in determining where responsibilities fall throughout the incident and what resources may be necessary to assist in response activities. A command meeting allows agency representatives the opportunity to develop a collaborative set of incident objectives and establish and agree upon a collective set of incident priorities.

These incident priorities and objectives are the basis of the IAP for the entire incident. During the command meeting, agency/jurisdictional representatives must also:

- Agree on the basic organizational structure

- Designate the best-qualified and acceptable operations section chief

- Agree on general staff personnel designations, as well as planning, logistical, and financial agreements and procedures

- Agree on the resource ordering process to be followed

- Agree on cost-sharing procedures

- Agree on procedures for the release of information

- Designate one agency official to act as the unified command spokesperson

It is important that each of these items is discussed and determined prior to the start of prolonged response activities, as they will provide a strong foundation for emergency responders.

Another feature of a unified command structure is that the incident command structure beneath the UC team is integrated and staffing can be appointed from any of the response agencies. At each level of the organizational structure, benefits exist in the staffing through multiple agencies/jurisdictions.

- Operations section. Within the operations section, staffing can be appointed from any of the response agencies. Deputy section chiefs may be utilized from other agencies to ensure that representation is even and that staffing is appropriately rotated to provide down time.

- Planning section. Within the planning section, staffing with representatives from multiple agencies/jurisdictions ensures that information is gathered from across all entities involved. It can also provide for a significant savings in personnel cost and time, allowing for agencies to rotate staffing into the section.

- Logistics section. One of the most beneficial uses of multi-agency/jurisdiction staffing is that it allows for an integrated communications system, maintaining a communication hub from one facility instead of multiple communication centers.

- Finance/administration section. Through cost sharing agreements, agencies and jurisdictions can ensure cost-effective operations are maintained and that the responsibility for provision of resources is appropriately divided among all response agencies (e.g., food, fuel, security, etc.).

With appropriate staffing assignments and management, the accessibility of a large pool of personnel will assist in the efficiency and effectiveness of every section and level within the command structure.

The final feature of the unified command structure is a coordinated resource request and purchasing process. Through the logistics and finance/administration sections, equipment, facilities, and personnel are requested or ordered through a single process. Each agency will utilize the same structure in making their requests. Those responsible for processing the requests can ensure that there is no duplication and that the requests are necessary and appropriate in the support of response operations.

Conclusion

Regardless of the call, the use of the appropriate ICS structure (single or unified command) provides an opportunity for effective, efficient, and appropriate management of the emergency responders. Becoming familiar with ICS through daily use will only help to strengthen your ability to utilize ICS in your operations at any emergency scene.

Scenario: Serving as EMS 1 this shift you have been lucky to have run a few calls around your area this shift, connecting with some old partners you had not interacted with in a while. While driving back to your station, the dispatcher asks you to call the dispatch center.

"Good afternoon, sir," this dispatcher states, "we have been requested to send an ambulance, engine, and an EMS supervisor to assist with a tactical operation. The SWAT team has to serve a warrant on a house. The team sergeant would like you to contact him immediately."

After hanging up the phone with the dispatcher, you call the sergeant. He advises you that they will be serving a warrant on a house that is a believed base of a methamphetamine operation. He asks that you participate in a planning meeting at 1700 hours to help coordinate details of the fire, EMS, and hazardous materials planning aspects of the operation. After hanging up the phone, you realize that while you can speak for the EMS aspects of planning, the on-duty battalion chief is more knowledgeable in the fire and hazmat planning, so you call and advise him of the meeting.

As you prepare for the planning meeting, you realize that this will become a unified response to the operations, with police serving as the lead, given the nature of the incident. You realize that you will be serving as the EMS agency IC and that you can expect to serve in the unified command post during the operation. With that in mind you begin to prepare the priorities for and possibilities of worst-case scenarios. After the planning meeting, you feel confident that you will be able to support the police operations and meet with the crew of Medic 1 to brief them on the upcoming incident. You recognize that they do not normally respond to a heavy police operation and want to ensure they understand that the command structure is still the same. After you provide them a briefing on the incident objectives, strategies, and tactics, you feel that they are prepared to assist in the call.

References

1. U.S. Department of Homeland Security, Federal Emergency Management Agency (FEMA) (March 2009). *Typed Resource Definitions—Emergency Medical Services Resources*. Retrieved March 9, 2011, from http://www.fema.gov/pdf/emergency/nims/508-3_emergency_medica_%20 services_%20resources.pdf.

2. FEMA (April 2005). *NIMS—Incident Command System for Emergency Medical Services, Instructor Manual*.

3. Limmer, D.D., Mistovich, J.J., Krost, W.S. (December 1, 2007). "Beyond the Basics: Scene Size-Up." Retrieved July 8, 2011, from *www.emsworld. com/print/EMS-World/Beyond-the-Basics—Scene-Size-Up/1$6754.*

4. International Association of Fire Chiefs (2008). *Fundamental Firefighter Skills*, 2nd Edition. Jones and Bartlett: Sudbury, MA.

5. fema (September 2005). "Unit 3: Basic Features of ICS." *ICS-100: Introduction to ICS – Student Manual*. Retrieved March 10, 2011, from http://training.fema.gov/EMIweb/IS/is100lst.asp.

6. FEMA (November 2004). *Introduction to Unified Command for Multiagency and Catastrophic Incidents*, 2nd Edition, 2nd printing.

7. FEMA (n.d.). "ICS 300—Lesson 4: Unified Command" [PDF document]. Retrieved March 10, 2011, from http://www.usda.gov/documents/ICS300Lesson04.pdf.

MASS CASUALTY INCIDENT

Chapter Objectives

Upon completion of this chapter, readers will be able to:

1. Understand the purpose and function of the medical group

2. Understand the difference between a medical group and medical branch

3. List the positions within the medical group

4. Understand the roles and responsibilities of personnel serving in the medical group

Scenario: While returning from the hospital from dropping off a medical patient you and your partner are discussing lunch plans when your tones drop.

"Engine 2, Squad 1, Medic 1, and Medic 2, you are responding to a motor vehicle collision. Caller states that a pickup truck and tour bus have collided and it appears there are multiple patients who have been ejected," the dispatcher tells you.

With all units marked en route, the dispatcher provides additional information.

"All units, be advised that the caller is on the line and advises that he can see flames in the truck and it appears there is someone still inside."

The first arriving fire units start requesting additional fire apparatus to assist with water supply due to the rural nature of the area. While responding, you also request that two additional ambulances be dispatched due to the number of potential patients.

What is your role in this incident? What needs to be done to ensure that all of the patients are taken care of? Do you have enough resources available to assist in handling the situation?

Introduction

What would you do if you encountered a situation like the one described in the above scenario? It is okay to admit the desire to put the ambulance in reverse and apply a large bolus of diesel fuel to remove yourself from the event. Most providers, evened seasoned, would admit to the same desire. However, as with your everyday calls, appropriate management of the scene can lead to an easy transition from start to finish while providing the best care to the patient.

A mass casualty incident (MCI) is any incident that creates more patients than there are providers to assist them. In the scenario above, the expectation is that a single ambulance will not be able to adequately care for the patients involved. As a result, a larger, more complex ICS structure is required. Because the primary focus of mitigation efforts for the incident is the triage, treatment, and transport of patients off the scene, EMS plays a crucial role in the response and recovery activities associated with an MCI.

Medical Branch vs. Medical Group

A single location MCI that involves an overall simple response provides the opportunity to organize the incident command structure in a set area, with limited extension outside the affected area. However, at times a mass casualty incident may extend to multiple areas, sometimes separated by a physical object or large land area. This can include highway incidents where the hot zone extends across both directions of travel, or a train incident where units approach from opposite sides of the train tracks. Figure 3–1 provides an example of such an incident. When this type of mass casualty occurs, multiple triage teams, treatment areas, and transportation groups may be necessary to ensure effective operations. In these instances, it is important to restructure the command system to ensure that the chain of command and span of control are within the appropriate limits. Establishing a medical branch will support this goal.[1]

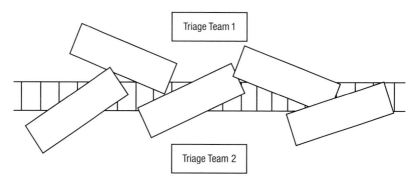

Fig. 3–1 This figure shows an incident that may require units to approach from both sides, with limited or no access to the opposite side of the incident.

The medical branch "is designed to provide the IC with a basic expandable system for handling any number of patients" at an MCI.[2] The medical branch is responsible for oversight of multiple medical groups set up to handle the patient care functions at the scene. The medical branch director is in charge of the medical branch. The director is responsible for ensuring that the strategy and tactics, as set out in the incident action plan, are successfully met. In an incident where development of the medical branch is necessary for successful operations, any established medical group falls directly beneath this position. In an incident such as the one illustrated in figure 3–1, a structure such as the one in figure 3–2 may be put into place.

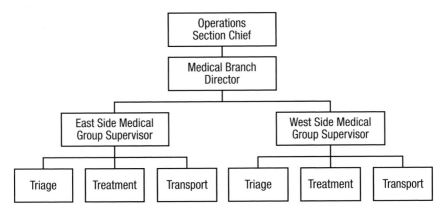

Fig. 3–2 This figure shows the ICS structure for an incident that may require multiple medical groups and the use of a medical branch director.

This structure ensures that patient care operations are completed in the most effective and efficient manner without significant duplication of activities.

Medical Group

When a mass casualty (multi-casualty) incident occurs, the ICS structure is expanded to include a medical group (figure 3–3). The medical group focuses specifically on the emergency medical service (EMS) efforts needed to mitigate a mass casualty/multi-casualty incident; in the ICS structure, the medical group falls under the operations section. Creation of a medical group allows for maintenance of span of control during the response to an MCI, most specifically when multiple medical groups are operating at one incident. In an incident where only one medical group is created, there is no need for the medical branch because the medical group supervisor takes on responsibility of all actions typically associated with the medical branch director.

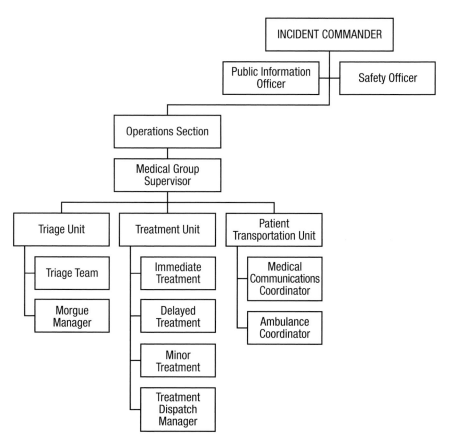

Fig. 3–3 Mass casualty incident command structure

The manager of the medical group is the medical group supervisor. This individual serves as the leader of all positions and activities within the branch and communicates directly with the operations chief, who is the direct supervisor of the medical group. As the primary supervisor of such a large group of individuals, the medical group supervisor has a large number of responsibilities. One major responsibility of the medical group supervisor is determining the priorities of actions for the staff of the medical group. While the overall scene activities are determined and finalized by the incident commander and/or unified command, the medical group supervisor ensures the appropriate tasks are assigned within the group. A second responsibility of the medical group supervisor is to set actions based on the tactics of the incident action plan (IAP). The incident action plan sets forth the tactics that are to be used to reach the ultimate goal of the command staff.

The medical group supervisor is also responsible for assigning resources to ensure completion of assigned tasks. Resource assignments should be made not only on the strategies set out in the IAP, but also on the capabilities of the resources available. Advanced life support (ALS) teams should be assigned to treatment areas, firefighters may be best suited for portering (movement of patients), and rescue truck personnel may best be utilized as members of the extrication teams. Additional responsibilities of the medical group supervisor include evaluating the progress of assigned resources to ensure they are appropriately completing their duties, and intervening and reassigning resources as necessary to ensure that the IAP goals are accomplished. As with all individuals on an incident scene, the medical group supervisor must maintain accountability of subordinates and ensure their safety is a priority.

Scenario: The officer of the first arriving fire apparatus assumes command of the incident and provides a size-up, confirming that the incident is multi-casualty. Upon your arrival, you are assigned the role of medical group supervisor. With this role, what assignments do you need to start making? What tasks need to be accomplished?

Components of the Medical Group

Scenario: As you arrive at the motor vehicle collision you see what, at this point, has only been words over the radio. A truck and tour bus collided in an intersection. The force of the collision has caused the bus to turn onto its side. Fire crews are working on extinguishing the truck fire and extricating the driver. Multiple patients appear to have either self extricated or were ejected from the bus, and a firefighter states that there are still a large number of victims on the bus. As you enter the scene the incident commander approaches you.

"I need you to get started in triaging the patients. We need to get a handle on how many patients there are. Let me know what resources you need so that we can get this task completed."

It suddenly hits you that you have just been assigned triage unit leader. Because no other ambulances have arrived on scene yet, you assign your driver and a firefighter the task of initial triage.

One of the most important tasks of an ICS supervisor is to choose the right people to complete a task. Triage teams may be most effective when staffed by basic life support (BLS) providers, as they are more likely to focus on the task at hand. ALS providers may feel a strong urge to use ALS skills to save a life, thus are best utilized in the treatment areas.

Triage unit

Personnel assigned to the triage unit are responsible for conducting and completing triage of all patients involved in the incident. The triage unit leader oversees the triage unit. This individual is responsible for oversight of all personnel assigned to the triage unit and ensuring that the tasks assigned by the incident action plan are complete. The triage unit leader must ensure that the triage of victims is completed in a timely manner and call for more resources as necessary. He or she must also decide the best method for ensuring completion of triage at a mass casualty incident and put that method into effect. The triage unit leader may choose the number of providers or teams of providers needed. Responders assigned to the triage unit are triage personnel. Triage personnel are responsible for the oversight and completion of patient triage. They must work in a timely manner, utilizing the appropriate triage method to ensure that all patients are triaged and accounted for. Triage personnel may then assist in the movement of patients from the incident site to the appropriate treatment area. Ensuring the appropriate number of triage personnel assists in effective and efficient completion of incident action plan tasks.

Triage systems. The triage unit is responsible for conducting and completing triage of all the victims that result from the incident. Triage is a method of separating victims to provide a better determination about those who are the most seriously injured and need more immediate assistance versus those patients who can be delayed in their immediate treatment. Triage also provides a better understanding of the severity of injuries and helps determine resources required for treatment, transport, and what facilities are needed to provide appropriate definitive care. While there are many triage system options available to EMS providers, some simple considerations should be kept in mind. The chosen triage system should be simple, easy to perform, and provide for rapid and simple life saving interventions. This allows for a quick review of a victim's injuries for rapid determination of triage priority. The chosen triage system should also be easy to teach and learn. Triage should be taught not only to the fire and EMS providers within a jurisdiction, but also police, dispatchers, and ER staff of the local hospitals. This ensures that all individuals involved in the first response to a mass casualty are working from the same triage system while mitigating the incident.[3]

The word triage has its origins in the French term "trier," meaning "to sort."[4] In its use in the emergency medical field, triage allows providers to sort patients into treatment categories to determine which patients are in need of more immediate medical assistance. While the earliest uses of triage can be traced back to the military battlefields, EMS providers may be more familiar with its uses in the Oklahoma City bombing, the terrorist attacks of 9/11/01, the London transit system bombings, or the Virginia Tech campus massacre. Each of these incidents created a situation in which the quantity of patients was much higher than available responders. As such, triage methods were enacted to assist in determining patient priority levels.

Every agency should select and train on a method or system of triage to be executed by its providers. The system chosen should meet certain criteria (no in-depth assessment, easily identifiable groups, easy to teach, easy to learn), but the specific system is much less important. The following text provide information on various triage systems utilized and available across the world. Regardless of the method used, it is important to remember that triage systems of neighboring jurisdictions should be similar so that providers can integrate appropriately into the operations of a scene.

Simple triage and rapid treatment (START). The START system was designed in the 1980s through a partnership between the Hoag Hospital system and the Newport Beach Fire Department in Newport Beach, California. The purpose of the system was to assist the hospital personnel in quickly organizing hospital resources in preparation for receiving victims from a mass casualty incident. Since its creation, the START system has become widely recognized and used (figure 3–4).

The primary goal of the START triage system is to "do the greatest good for the greatest number." START is designed to allow providers to assess victims in less than one minute to determine the severity of injuries and need for additional medical care. During assessment of each victim, the primary focus is assessment of respirations, pulse, mental status, and ability to walk. Based on those findings, patients can be appropriately categorized.

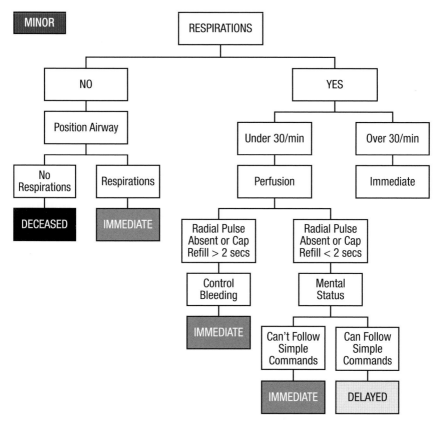

Fig. 3–4 START triage algorithm

Triage of patients using the START system focuses on categorizing patients into four groupings using colors to differentiate between the priorities. The first category is "immediate," with patients tagged red. Immediate patients experience problems with respiration, perfusion, and mental status. They require immediate medical attention. These patients are first to be transported from the incident. The second category is "delayed," with patients tagged yellow. Delayed patients often suffer from burns without airway problems, major or multiple bone or joint injuries, or back or spinal injuries. These patients will survive if definitive medical care is not received immediately. The third category is "minor," with patients tagged green. Minor patients are those patients often considered the "walking wounded." They may suffer from cuts, scrapes, and sprains, and can respond to the request to walk from the scene. These patients would often be considered for refusals if they were not part of a mass casualty. The fourth category is "deceased/non-salvageable," with patients tagged black. These patients are not breathing and while they would be resuscitated in a "normal" situation, the resources necessary for that effort are not available.

One of the most significant strengths in use of the START system is the fact that assessment of patients is designed to be completed in less than one minute. The ability to complete primary triage so quickly allows the provider to move from patient to patient quickly, in order to "do the greatest good for the greatest number." Another positive aspect of the START method is that it was designed to integrate into the ICS structure. This ensures that departments that use the START methodology can maintain NIMS compliance while working to manage mass casualty incidents.

One recognized weakness of the START system is the secondary triage system. While secondary triage does occur, it normally occurs prior to any treatments. Therefore, secondary triage does not take into consideration any improvement in patient conditions because of treatments received.

Sacco triage method. Created by Dr. Bill Sacco and based on an analysis of over 100,000 patients, the Sacco triage method (STM) focuses on not only patient condition, but also the number of resources available for management of the mass casualty.

The primary goal of the STM is to maximize the expected number of survivors. This is done by using an evidence-based, outcome-driven method designed to be used on a daily basis as well as during a mass casualty incident. Categorization using the STM focuses on respirations, pulse, and motor response. Each assessment is scored on a scale of 1 (one) to 4 (four), and scores are combined to determine the category of the victim.

The Sacco triage method does not specifically designate categories with colors or terms. Categories are created based on the assessment scores that are assigned to each patient. Patients who receive a score of 0 (zero) display no signs of life, and patients who receive a score of 12 are minimally injured or not injured at all. Patients who receive lower scores are considered in need of the highest level of care and are transported first.

One of the major strengths of the Sacco triage method is that it takes into consideration all regional resources in making the triage and transportation decision. The computer software that is utilized monitors both patient categorization as well as the available number of transport vehicles in determining which patients to transport from the scene at any given time. Another aspect of the STM that makes it beneficial to departments that utilize it is the ability to use the triage method on everyday patients. The constant use of the triage protocol ensures that you, your fellow firefighters, and neighboring agencies are familiar with the system so that when you need to use it in a mass casualty it is second nature. As with other triage methods, the STM allows for reassessment of patients to upgrade or downgrade their triage categorization based on their response to treatments and other factors.

While the software used in the STM can benefit patients by assisting in the transport decision, it can also be a disadvantage. Because the Sacco triage method relies on a computer system to make the transportation decision, departments must purchase the software to be able to use the STM as it was designed. This may prove cost prohibitive for many departments, as it would involve assuring availability of computers for the field and any supporting equipment needed. Another negative aspect of the Sacco triage method is that the forms utilized in the system are not NIMS specific forms. If your locality chooses to use the Sacco triage method, you may encounter problems working in mutual aid response situations with localities that use the standard NIMS forms.

Triage sieve and sort. The triage sieve and sort method is a triage system used most often in the United Kingdom to assist in mass casualty management. The system is designed to provide primary and secondary triage. The primary triage is the triage sieve, during which providers quickly sort casualties into groups based on priority of treatment need. The second step, triage sort, is a more in-depth assessment and sorting of patients based on not only injuries, but also available resources for on-scene treatment and transport.

The methodology of triage sieve and sort is simple. With the triage sieve, what appears to be the standard triage process is utilized to assessment patients. Assessment of the injured victims is first based on the ability to walk. From there, assessment continues with assessment of airway and adequate breathing. However, unlike other methods such as START, the triage system does not assess presence of a radial pulse. Instead, the triage sieve and sort method relies on the assessment of capillary refill to complete the triage of each patient. Another difference in the methodology of the triage sieve and sort method is that after assessment is complete, patients are categorized based on the intensity of interventions required to assist the patient and not on their specific injuries.

As with other triage systems, the triage sieve and sort method places patients into four categories. The first category is P1 and is what other systems refer to as the "immediate" category. Patients in this category will die without life saving interventions. This may include airway placement, breathing assistance, and major bleed treatment. These patients are identified by the color red. The second category is P2, or "intermediate," and as with other systems, uses the color yellow to identify these patients. Patients in this category require a significant number of interventions but can survive a few hours without the interventions. The third category is P3, "delayed" and identifies patients using the color green. These patients may have walked away from the scene when originally asked, or may have minor injuries such as cuts and scrapes. The last category of the triage sieve and sort system is for those patients who are non-salvageable. This category is referred to as "dead," and as with the START system, designates these patients with the color black.

One of the strengths of the triage sieve and sort system is that once past the initial triage phase, there are additional categories that can be used to further prioritize patients for treatment and transport. As with START and the move, assess, sort, and send system

(MASS, discussed below), the triage sieve and sort system uses a fifth category to designate those patients who without immediate assistance would be unlikely to survive their injuries. This ensures that those people with the highest probability of survival get treatment. A second strength of the triage sieve and sort system is the tagging system. Use of a tagging system ensures that a patient is tracked from beginning to end, documenting initial triage, secondary triage, treatments conducted, and personal information of the patient. This documentation ensures that all patients are tracked and that continuity of care is maintained throughout the event.

In most triage systems, the secondary triage assessment is normally a strength. In the triage sieve and sort system, secondary triage is a weakness. While initial triage of the system is based solely on physiological information (respirations, pulse, etc.), the secondary triage is based on each provider's knowledge of injuries and their effects on the anatomical structure of the patient. An advanced trained provider may be able to retriage based on a more specific and better thought out assessment of the patient than a provider with the basic knowledge of emergency medical care. This use of personal knowledge removes the standardization of triage and does not ensure that all patients are being triaged in the same manner.

Move, assess, sort, send (MASS). The MASS method of triage is loosely based on the triage system designed for use in the military setting; however, modifications were made in its design for use in mass casualty incidents. The MASS triage system is taught as a part of the basic and advanced disaster life support (BDLS and ADLS) courses offered throughout the country.

The methodology of the MASS system is that each letter of the acronym stands for a step in the process. The first step is move. As with START, you relocate the victims who can move to a designated location outside the hot zone. Unlike the START method though, victims are instructed to move a limb. Based on the patients who do not move a limb when asked, the provider moves to the second stage, assess. During the assess stage, the triage team focuses on the patients who do not move limbs and assesses them using the standard assessment of airway, breathing, and circulation. Based on assessment findings on these patients, the triage team moves into the sort stage. Once assessed, patients are sorted into one of four categories and the final step of the MASS triage system, send, is carried out. At this point, patients are sent to hospitals for additional treatment.

As with the START triage system, the MASS system places patients into four categories based on the need for immediate medical treatment. The categories used to accomplish the sorting of patients are referred to as "ID-me," which stands for immediate, delayed, minimal, and expectant. Each category is based on the injuries presented. Patients with no airway, respirations, or circulation are placed in the "expectant" category. Patients who were able to move from the scene when instructed by the rescuer are placed in the "minimal" category. Patients who were able to move limbs when directed to by the rescuer are placed in the "delayed" category. Patients who could not move limbs when requested to and have open airways are placed in the "immediate" category.

As with the previously discussed triage methods, the MASS system allows for additional triage of the immediate patients. After the initial categorization of all immediate patients, additional sorting of the patients based on which patients are most critical is completed, if there is time. Patients who are more likely not to survive, even with additional treatment, are placed in another category that places their treatment and transport priority after immediate patients and before delayed patients. A second strength of the MASS triage system is that it allows civilian and military responders to interact at the scene of major emergencies. Because the MASS system is based on the military triage system, civilian responders will be able to interact with military providers when working together to respond to a major incident. This allows for a smoother transition from military to civilian and civilian to military responders.

One of the most significant weaknesses of the MASS triage system is that it relies on the visibility of patients to determine initial triage priority. At an incident scene where you do not have visibility of all patients or forget which patients could follow commands, you will have to repeatedly conduct the first step of the MASS system with each patient. This repetition will only lengthen the time it takes to triage all patients. A second weakness of the MASS system is that reprioritization of patients is based not on treatment response but on the number of patients to be transported in a higher priority group. For instance, if there are only two immediate patients to be transported and twelve delayed patients, then the delayed patients receive a high priority during secondary triage, regardless of whether they are in need of it or not. A third weakness is that unlike some other triage methods, the MASS system does not incorporate a tagging system to identify patients, their priority, and any treatments that may have been conducted. Fire and EMS systems that utilize the MASS triage method must also select and train on a tagging system that may or may not fully integrate with the MASS system methodology.

SALT. SALT triage is a movement toward a national standard of triage. SALT is the acronym for sort, assess, life-saving interventions, and treatment and/or transport. The SALT triage process begins with a verbal request that all patients who are able to move do so. This provides a global sort, which allows you to then determine which patients should be assessed first. Patients who made no purposeful movements or who have obvious injuries are assessed first, patients who waved or made purposeful movement are assessed second, and those who walk to the designated location are assessed third. From this point you conduct individual patient assessments, sorting them into the appropriate triage category. As with a majority of the triage systems used, the SALT method utilizes the four major categories: "minor," "delayed," "immediate," and "dead." However, unlike other methods, SALT triage recognizes that some patients may have significant injuries that may indicate survivability is in doubt. These patients are categorized as "expectant." As with each triage method, there are numerous pros and cons that should be reviewed when determining which triage method is best for your agency.

Careflight. The careflight triage method is one of a few triage systems recognized by and used in the Australian fire and EMS system. As with the START, MASS, and triage sieve and sort methods, the careflight triage method assesses a patient's ability to walk as the first step. From this first step, the careflight method also mirrors the START method by focusing on a patient's airway and breathing status, presence or absence of a radial pulse, and the ability to follow commands to complete the triage of all victims of the incident.

Reverse triage. From its originations in the military, triage systems have been modified to meet the needs on the battlefield. With needs in the battlefield constantly changing, a system of reverse triage was created. Reverse triage focuses on the need to treat those with the most minor injuries first. This allows for the return of personnel into the field. In the military setting, reverse triage is used to return soldiers with minor wounds back to the battlefield so that they can continue to support the fight. For fire and EMS providers, reverse triage may be necessary in a situation where a large number of medical providers are among those injured and treating their minor injuries will allow them to return to the scene to provide medical care. While not a standard triage method, the reverse triage method may be useful to emergency responders. In terrorism preparedness courses, first responders are warned that they can become secondary targets in terrorist events. Should a secondary attack occur, reverse triage would allow more responders to continue helping civilian victims.

Tracking your patients. Tracking of patients is a key element of mass casualty incident management. It allows personnel to understand the number of patients involved in the incident, the resources needed to treat these patients, and allows receiving facilities to prepare for an influx of patients. Patient tracking usually begins with initial triage. Marking or tagging patients from the first encounter allows you to track them either by color or by patient number. When the triage teams begin conducting triage they should have appropriate resources to mark their patients. There are two main methods for marking patients, one used during initial triage and the other utilized to gather more information on the patient while tracking his or her condition.

Triage tape. The use of triage tape allows the personnel conducting triage to quickly mark a patient with the appropriate color before moving on to the next patient. The type of tape used is not important, and can be simple land surveyor's tape, which comes in the four colors of the triage categories (as well as many other colors). The tape is a simple way to mark triage categories quickly while moving through an incident scene and is usually secured to the patient by tying it around a limb in a visible place. Tagging patients allows personnel to count them, categorize them, and also provides a visual marker to those who may later provide aid to the patient. For instance, when porters are tasked with moving patients from the hot zone to the treatment area, they will rely on patient tagging to move priority patients first.

Triage tags. While tape is excellent in providing initial markings of triage categories for patients, it does not provide any other information. After patients are triaged and moved from the incident site to the treatment area, additional tracking mechanisms should be utilized, both to track the patient and to ensure that patient information and treatments are gathered and documented throughout the incident. In the treatment area, when secondary triage is conducted a triage tag should be applied. A triage tag serves as the patient care report in a mass casualty incident. While there are many triage tags available commercially, there are a few key items to consider when determining which triage tag to use. The first is that it should allow for the documentation of patient information (i.e., name, address, sex, etc.). It should also have an area to document multiple triage outcomes to indicate the injuries, though it can only include general body areas and types of injuries. You may also want to ensure that your tag allows for documentation of decontamination status. This is most helpful in a hazardous materials situation. Effective triage tags should also include an area to document assessment findings and corresponding treatments. Most importantly, a triage tag should be numbered. Providing a unique number to each triage tag assists in tracking a patient from triage tag placement to his or her final destination.

Morgue manager. Immediately after determining the incident has created even one deceased patient, the triage unit leader must recognize that additional personnel will be needed and immediate actions must be taken. At the determination that individuals have died because of the incident, the triage unit leader should follow local protocols and ensure notification of the appropriate personnel to begin the investigation. In some localities, this may be the chief medical examiner, while in others it may be the on-call funeral director. The triage unit leader must also designate a morgue manager. The morgue manager is responsible for establishing the morgue area and providing oversight of this area until relieved. The morgue manager coordinates with law enforcement and the coroner or medical examiner representative to ensure that all deceased persons are appropriately handled and moved when authorized. Keep in mind that movement of patients prior to authorization should not occur unless it is necessary to access a salvageable patient, as the deceased persons are considered evidence. If you are appointed morgue manager, you may also find yourself coordinating movement of deceased patients with the treatment unit leader should a patient die in one of the treatment areas.

Treatment unit

Scenario: Assume that instead of triage unit leader you are assigned the role of treatment unit leader at the motor vehicle collision described at the beginning of this chapter. The triage unit leader reports that you have 20 minor, 10 immediate, 15 delayed, and 5 deceased. You recognize the need to establish treatment areas and begin work finding the appropriate locations and making them visible.

"Operations from treatment unit leader. I need additional resources to assist in staffing treatment areas. Please send me two ALS providers and one basic provider. I will also need 12 road cones to assist in setting up the treatment area."

"Operations copies. I will get you those personnel as soon as possible."

With the personnel on the way, you begin work on designating the appropriate areas based on space. Given the numbers provided you know that delayed and immediate treatment areas will need to be fairly large. You also begin considering the other resources you are in need of to ensure the best care is provided to the victims.

What are the best locations for the treatment areas that are safe for everyone? What supplies do you need to appropriately and adequately treat the patients? Where will you get the supplies? Will you need additional personnel to assist in treating patients? What steps will you take to ensure an accurate patient count is maintained?

Once triaged, patients are moved from the hot zone, the area of the incident, to the treatment area. The treatment area is where patients are placed, monitored, and given appropriate aid until transport to a medical facility can be provided.

The treatment unit leader oversees activities in the treatment area and is directly responsible for all activities related to the treatment of patients at the mass casualty incident. He or she must ensure that patients are appropriately distributed and an adequate number of treatment personnel are within the area to assist with procedures and care. Not only does the treatment unit leader interact with the triage unit leader to determine how many patients will be entering treatment, but is also responsible for working directly with the patient transportation unit leader to ensure that the right patient is transported at the right time to the right facility.

If you are assigned the role of treatment unit leader, you should consider staffing the following positions to assist in completion of incident objectives

Treatment dispatch manager. The treatment dispatch manager is responsible for the coordination of movement of patients from treatment to patient transportation. The treatment dispatch manager should consider the prioritization of patient needs when determining which patients are transported first. He or she must also work with the medical communications coordinator to ensure that appropriate units are being used for transport (advanced life support, basic life support, aeromedical, or ground unit). This position must also maintain patient accountability to ensure that the disposition of each patient is appropriately documented.

Treatment area managers. Because of the complexity of treatments needed for the various levels of triage, treatment areas should be divided into three areas: immediate, delayed, and minor.

Immediate treatment area manager. Patients in the immediate treatment area have the most significant life-threatening injuries, usually problems associated with respirations, perfusion, or mental status. The immediate treatment area manager provides oversight to patients and providers in the immediate treatment area. Because of the injuries associated with "immediate" patients, highly trained providers are the most appropriate personnel to staff the immediate treatment area. In large incidents, trained personnel may include not only paramedics and other ALS trained providers, but also physician assistants, doctors, etc. There is also a need for significant amounts of patient care equipment in the immediate treatment area, including monitors, oxygen, suction, etc. The immediate treatment area manager must also ensure that patients in this area are appropriately prioritized for transport to off-site medical facilities.

Personnel in the immediate treatment area will face large amounts of stress as their patients are the most critical with the most significant injuries. Activities must be quick, as a breakdown between treatment and transport increases the possibility of loss of life. Because of the potential for large number of patients moving quickly through the immediate treatment area into transport, any personnel stationed to work in the area should frequently rotate in an attempt to minimize the effects of the stress.

Delayed treatment area manager. The delayed treatment area is used for patients who are less critical and can tolerate a delayed treatment. The delayed treatment area manager provides oversight to patients and personnel in the delayed treatment area. This position ensures that patients receive the appropriate care and that they are retriaged as necessary, coordinating movement to the appropriate treatment area if needed. Patients who move from the incident area into the delayed treatment area tend to have minor problems with airway. Within the delayed treatment area, the focus is on preventing shock, respiratory problems, and stabilizing any fractures prior to movement to the transportation area. The delayed treatment area manager must also ensure that patients in this area are appropriately prioritized for transport to off-site medical facilities.

Minor treatment area manager. The minor treatment area manager provides oversight to patients and personnel in the minor treatment area. These patients present with minimal injuries and are initially alert, oriented, and walking. Because of the nature of the injuries within this group, patients within the minor treatment area are typically transported last, if at all. These patients may actually have injuries; however, any individual involved in the incident should be accounted for, from rescuers to victims. In the scenario in this chapter, you would account for all victims, including the passenger who was able to escape uninjured and call for assistance. Because he might show problems later, you want to be able to ensure that you include him in your continued assessments.

Medical supply area. One consideration during a mass casualty incident is the need for a medical supply area. The medical supply area allows for a centralized placement of EMS and medical equipment within the treatment area so personnel have easy access to the necessary treatment supplies. The medical supply area is supervised by the medical supply coordinator. This individual is responsible for management of receipt and inventory of equipment in the area. There are multiple layout options for a medical supply area. One layout would be to place medical equipment in one central location for all treatment areas to access and utilize. Figure 3–5 illustrates this possible layout. While this ensures that everyone has access to the medical equipment, you must consider the location of the medical supply area to ensure that each treatment area can access it quickly. A second layout option for the medical supply area is to create a small area for storage of medical equipment for each treatment area. Figure 3–6 illustrates this layout. While this ensures that each treatment area has quick and easy access to the equipment, consideration must be given to ensure that adequate equipment is available in each area. In separating medical equipment, you should also keep in mind what specific equipment each treatment area may need. For example, the red treatment area will need more invasive equipment such as IVs, oxygen, etc., while the minor treatment area may only need equipment for treating minor cuts, scrapes, and bruises. An additional consideration when setting up the medical supply area for the treatment areas is equipment familiarity. During mutual aid response, equipment may not be similar to that normally used, and providers may be unfamiliar with it.

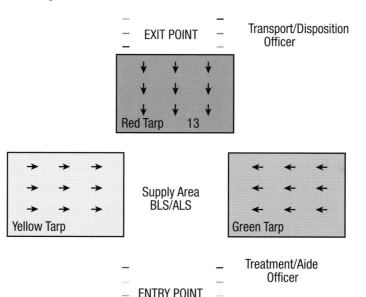

Fig. 3–5 This figure shows a treatment area layout that provides for a single cache of equipment for all treatment areas.

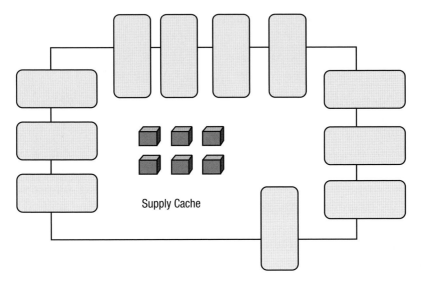

Supply Cache

Fig. 3–6 This figure shows an example of a treatment area layout that provides a cache of equipment for each treatment area.

Setting up the treatment areas

Even before triage is complete the setup of the treatment areas needs to begin. However, keep the following items in mind when determining location, size, and logistics associated with an effective treatment area.

First, the treatment area needs to be in a location that allows for expansion and reduction in the size of each area. When first established, the treatment area size will be based on an initial estimated number of patients. Because this number can change as providers complete triage as well as conduct secondary triage, space should be adequate to accommodate a large (or small) number of patients. The space should also allow for providers to be able to move between patients and treat patients as necessary. This can be best accomplished by considering each patient laying on a backboard. A backboard is 6 feet long and approximately 24 inches wide, and you should allot enough space for a backboard as well as for walking room around the patient. This means that each patient should be allotted a space approximately 7 feet long by 3 feet wide. This will allow enough space to assess the patient and move around to all patients in the treatment area.

The next consideration in setting up a treatment area is weather conditions. You should consider not just the conditions at the time of arrival on scene, but also the conditions that may occur as the event progresses. In the scenario at the beginning of this chapter, the weather is nice, 75 degrees and partly cloudy with a low humidity. However, a thunderstorm or other weather event may be on the horizon, posing a threat to the victims of the car accident. With this in mind, consider your ability to protect your patient from weather issues, including wind, rain, snow, sleet, cold, and heat. One idea for protection is pop-up shelters. When put together, these may provide shade and protection from some weather issues. However, an extremely large incident may require numerous tents, which can become costly and cumbersome for storage when not in use. Another idea is using buildings that may be within the incident area. For example, an incident at an after-school football game may require the use of the school facilities such as the cafeteria or gym for the placement of patients. Weather protection may also be as simple as providing a blanket and/or sheet to assist in the warming of the patient if other protection is not available.

A third consideration in the setup of the treatment area is the ability to make the area noticeable to the on-scene responders. Once you have determined how and where you are going to set up the treatment area, you need to make sure that all responders on the scene, more specifically those who may need to transport patients into or out of the treatment area, can find the area and the specific treatment areas within it. One way to mark the area is with ground tarps colored to specify the treatment area color (i.e., immediate, delayed, minor, or deceased). This would allow for a visual marking of the area. Keep in mind that a ground tarp is quickly covered by patients once they are moved from the incident area to the specific treatment area. It also requires the providers look at the ground, creating a safety hazard as they move patients. A second way to mark the treatment area is with colored pop-up tents. These tents provide both a visual marking for the providers as well as weather protection for the patients. While these provide dual functions, as mentioned earlier, an adequate number of pop-up tents may become financially impossible for your agency to secure, as well as cumbersome to keep stored and transport to an incident. A final potential method for marking the specific treatment areas is the use of colored flags. Colored flags placed on poles or posts are easier to secure in the line of sight of the responders. They allow for simple marking and provide a quick visual to ensure that patients are placed in the appropriate treatment area. Colored flags are less expensive and easier to store then some of the other marking alternatives.

Patient transportation unit

Scenario: The operations section chief has assigned you the position of patient transportation unit leader.

"Be advised triage has been completed and they are working to move patients from the incident site to the treatment areas," he tells you.

While not exactly eager, you accept the position and begin to assess the situation.

"Patient transportation from treatment be advised, we are beginning to receive patients and would like to begin transporting them as soon as possible."

It seems like a simple request, but having just received the position of patient transportation unit leader you are already a bit overwhelmed.

What resources are available for you to use for patient transport? What facilities have agreed to receive patients from the incident? Who do you need to coordinate with to ensure the appropriate patient is transported at the appropriate time?

With both triage and treatment handled, transportation should become the next focus. In a mass casualty incident, the patient transportation unit leader (or group supervisor) is assigned to the medical group and reports directly to the medical group supervisor. The patient transportation unit leader coordinates all aspects of moving patients from the scene to additional medical care. This may include area hospitals or medical centers that will assist in more specified treatment as necessary. A patient transportation unit leader provides oversight to the transportation unit. This individual should have knowledge of air and ground ambulances within the area and have a background in resource management. The patient transportation unit leader is responsible for the coordination of all transportation resources, both ground and air, utilized for the movement of patients to medical facilities, as well as the maintenance of transportation records and coordination of patient movement from treatment to the transport area.

Transportation vehicles. Ensuring that patients are quickly transported and in an appropriate manner is essential in the management of a mass casualty. Ambulances are the most desired and easily accessible modes of transportation for a mass casualty. They provide not only the equipment for the medical supply area, but also a vehicle to move patients. However, a standard ambulance can only transport two patients on backboards, minimizing the number of patients that are moved from the scene at a time. Another disadvantage is the aspect of transport time. Ground transportation is slower than air transportation. At a mass casualty, you should consider not only the transport time to the hospital, but also the time it will take the crew to transfer patient care to the hospital staff and restock equipment. This can add upwards of 30 minutes to the time spent away from the scene.

Air transportation (usually via helicopter) provides another alternative for moving patients from the incident scene to medical facilities. While they are limited in numbers in some areas, there are significant positives to their use during a mass casualty. One positive is that they allow you to transport patients to medical facilities that are farther away from the incident site, so that ground units can transport patients to facilities within a short driving distance. They can also ensure a quick transport time for the most critical patients, allowing those critical patients to receive advanced medical care more immediately than if they were transported by a ground unit.

When using air transportation, keep in mind the following considerations. First is that you should not rely on the ability to use a helicopter. Weather, mechanical problems, and use in other incidents may limit your ability to utilize aeromedical transport. However, if units are available, you should utilize standard dispatch procedures to request the assistance of the helicopters. Notification should also be made as early as possible. In the scenario discussed throughout this chapter, you would want to request the assistance of a helicopter upon arrival and recognition of the large number of patients.

Once the request for a helicopter has been made, actions should be taken to set up a helispot. The helispot should be in an open area and meet the needs of the incoming equipment. This may include sufficient size for multiple helicopters to land. The area should be clear of debris and hazards including overhead wires, as well as appropriately lit for easy visibility. One thing you should do in the preplanning/preparation stage is schedule meetings with area helicopter services to determine what their requirements are for a helispot. Additional resources may also be necessary to support helispot operations. First, most helispots require the presence of an engine or some type of water source, should the helicopter experience problems during landing or takeoff. This means that an additional crew and additional equipment will be necessary to support the helispot operations. Another consideration is the proximity of the helispot. If it is a too close to the scene or treatment area you will need to ensure that patients will be protected as the helicopters land and take off. If a helispot has to be set up a distance from the scene, then transport of the patients from treatment to the helispot is a major concern. If you are serving as patient transportation unit leader, you will need to work directly with other members of the ICS structure to ensure that you are coordinating efforts of patient transport with the air ambulance coordinator.

Staging. With the potential for a large number of transportation vehicles, management of the vehicles upon their arrival is a necessity. The first step to manage incoming units should be from the first arriving unit. Upon providing a scene size-up and verbalizing command, the first arriving officer should establish a location for all incoming units to stage. The staging area is beneficial as it allows the patient transportation unit leader the ability to know what pool of resources is available for transporting patients from the scene. However, until an individual is assigned the task of managing staging, the patient transportation unit leader is responsible for calling for each resource. In order to ensure that transportation decisions are made efficiently, an ambulance coordinator should be quickly assigned.

The ambulance coordinator works to establish the staging area for all patient transportation vehicles (ground and air), maintains records of ambulances that have been assigned a task and are available for assignment, and also works with the transportation unit to ensure that the right unit is assigned for each patient. The ambulance coordinator must also ensure that ambulances used for transport are provided with information on safe and open routes to the medical facilities and that directions are available for those personnel who may not know the location of receiving facilities (figure 3–7). The ambulance coordinator also works with the logistics section to provide personnel with a list of medical supplies that are available from the transporting units. These supplies may be used for the medical supply areas for the treatment unit.

In order to ensure that the transportation unit runs efficiently and maintains the appropriate span of control, additional roles are necessary. The first role is that of loader. As patient transportation unit leader, you need to assign at a minimum two people (four or more may be most beneficial) to serve as loaders. Loaders are individuals who move patients from treatment to the ambulance that will be transporting. By assigning specific individuals to fill this role, patient movement and removal from the incident scene can occur in a smoother manner, with fewer holdups because of insufficient personnel. By assigning a larger number of personnel to the role of loader, you can also ensure that they do not tire as quickly and can serve in the role for a longer duration.

Fig. 3–7 The staging area should be large enough to hold all dispatched units

An additional position that may assist the patient transportation unit leader is the medical communications coordinator. This individual serves as an on-scene dispatcher of sorts. The medical communications coordinator interacts with the hospitals that will be receiving patients and with the ambulance coordinator to determine which ambulances will be loaded for transport at what point. This person works as part of the chain to ensure that the RIGHT patient is taken to the RIGHT hospital with the RIGHT mode of transportation. The medical communications coordinator also works with the air operations branch director, who provides coordination and oversight of the aeromedical resources. The medical communications coordinator requests air ambulance transportation as determined by the treatment dispatch manager (figure 3–8). Figure 3–9 provides a visual guide to how communications should flow during a mass casualty when moving patients from the incident to the hospital.

Fig. 3–8 Air transportation provides another alternative for moving patients from the incident scene to medical facilities

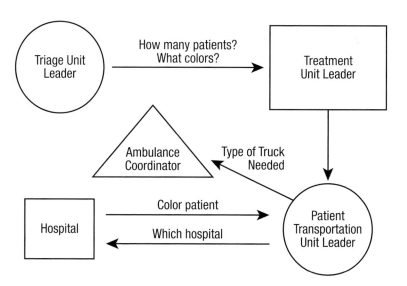

Fig. 3–9 This figure provides a flow chart for communication during a mass casualty as it relates to patient movement from the hot zone to a definitive treatment facility

Scenario: You have finished your daily check sheets on the medic and are preparing to sit down for a cup of coffee when suddenly the claxon goes off and the dispatcher begins to talk.

"Engine 1, Engine 2, Truck 1, Medic 1, Medic 2, and Battalion 1, respond to Pine High School for a roof collapse."

Initial information states that the damage was in the cafeteria area and there are injuries. As you mark en route you begin to do a mental checklist of what you need to do upon arrival. With the recognized potential for multiple injuries, you request three additional ambulances.

"Battalion 1 on scene. I have a large single-story brick structure with visible damage on the A/B corner. Battalion 1 will be establishing Pine High School command. Engine and Truck 1, from command, assess the structure for stability and perform a primary search to identify any potential victims. Medic 1, upon your arrival begin setting up the medical group."

Your unit and Medic 2 arrive on scene at the same time. You assign the crew of Medic 2 to triage and begin the steps to set up treatment, transport, and acquire the necessary resources to help all of the injured.

"Medical group director from triage."

"Go ahead."

"We have completed triage. We have six immediate, four delayed, eight minor, and two deceased."

"Understood, triage group. Please report to staging for reassignment."

As you receive this radio traffic, Medic 3, 4, and 5 arrive on scene. With the completion of triage, you reassign Medic 2 to rehab and advise them upon completion of rehab they are to return to staging. You also hear from operations that additional personnel have been placed in staging after completing their initial taskings and participating in rehab. You realize that there is a need for adequate resource allocation for the treatment areas and make the request to the operations manager for an engine crew (all EMS certi-fied) to report to you. After checking with the staging manager, the requested resources are assigned. With the additional resources, you then make assignments.

"Engine 2 and Medic 3, report to the treatment unit."

When the crews from Engine 2 (four FF/EMTs) and Medic 3 (paramedic and EMT) arrive at the treatment area, the Engine 2 officer is assigned as the treatment unit leader, and with this begins making assignments. With the role of treatment assigned, you begin focusing on the transportation needs.

"Medic 4 from medical group director."

"Medic 4 officer, go ahead."

"I am assigning you the role of transportation unit leader. Meet me face to face at the command post for further instructions."

"Medic 4 understood."

After briefing the OIC of Medic 4 you take a second and deter-mine that it is a good time to review the incident action plan (IAP) and determine if the objectives set in the IAP are being met. You begin with a review of the treatment area (figure 3–10).

Fig. 3–10 Use of colored tarps is one way to mark treatment areas, but the size of the tarp can be a limitation to its use.

The treatment officer provides you the following update:

- All patients, except black, are now in the appropriate treatment areas. Black patients are still in their original locations, as a medical examiner has not stated that the victims can be moved.

- Two members of Engine 2 have been assigned to a BLS treatment team. The third member of Engine 2 is assigned the position of minor treatment area leader.

- The paramedic from Medic 3 is serving as the immediate treatment area leader and the EMT is serving as the delayed treatment area leader.

- There is a need for an additional person to serve as the morgue manager.

With this update, you advise the treatment officer to contact staging to obtain the necessary personnel. You also ensure that they have the necessary supplies to continue treatment until the patients are transported and remind the treatment officer to contact you if there are additional needs.

Feeling satisfied that the treatment unit is continuing to meet the expectations of the IAP, you check in with the transportation officer. He provides you with the following report:

- He has assigned his partner the position of medical communications coordinator.

- Hospitals have been notified of the ongoing incident and are gathering a bed count to relay back to the medical communications coordinator.

- An aeromedical unit needs to be requested and a helispot established.

Recognizing that the transportation area deficits will create a log jam and increase the on-scene time of the patients, you begin to work with the patient transportation unit leader to find solutions to the issues (figure 3–11).

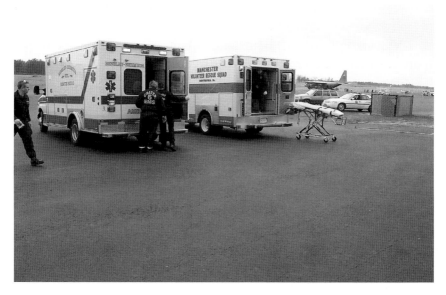

Fig. 3–11 The transportation officer should ensure that transport units are have access and egress to pick up patients from the treatment area.

After some discussion, the following actions are taken:

- The transportation officer contacts staging and assigns Engine 1 the task of setting up a helispot.

- You contact operations and request that three medics and one air ambulance unit be dispatched to staging/helispot.

- You assign one of the firefighters from Engine 1 to the position of transportation recorder to ensure that the appropriate documentation is maintained.

As the incident progresses you continue to monitor the situation to ensure that the incident objectives are being met and that no portion of the medical group goes without assistance.

"Transportation from treatment, be advised you have received the last patient."

"Transportation understood."

"Medical group leader from treatment, be advised there are no more patients in treatment and I am releasing personnel to rehab."

After acknowledging the radio traffic, you pass the information to the operations section. With the loading of the last patient, you have successfully transported all patients to definitive medical care and can begin to "wrap up" your activities as they relate to the response phase of this operation.

Conclusion

The scenarios presented in this chapter represent only a small portion of the potential scenarios that could that generate a mass casualty incident. As an EMS provider there are many roles that you can fill within the mass casualty incident command structure, outside of the basic ICS roles. Each of these roles carries with it specific roles and responsibilities. However, whether you serve in a leadership role within the ICS operational structure or you just drive an ambulance to transport patients to the hospital, your role is important in the mitigation of a mass casualty incident.

References

1. U.S. Department of Homeland Security, Federal Emergency Management Agency (FEMA) (April 2005). *NIMS: Incident Command System for Emergency Medical Services, Instructor Manual.*

2. Christen, H. & Maniscalco, P.M. (2002). *EMS Incident Management System: EMS Operations for Mass Casualty and High Impact Events.* Prentice Hall: New Jersey.

3. OES Fire and Rescue Service Advisory Committee/FIRESCOPE Board of Directors (December 1989). *Multi-Casualty Operational System Description*, ICS-MC-120-1. Governor's Office of Emergency Services: Riverside, California.

4. "Triage." Retrieved March 10, 2011, from www.merriam-webster.com/dictionary/triage.

DISASTER RESPONSE

Chapter Objectives

Upon completion of this chapter, readers will:

1. Understand the process for requesting assistance during a major incident

2. Understand the different levels of resource assistance

3. Understand emergency operations center (EOC) operations and their role in disaster response

4. Understand the ICS and EOC interface during disaster response

5. List available EMS resources at the local, state, and federal level

Scenario: Walking into your station to start your duty crew, you sit down just in time to hear the weatherman give updated information on the potential threat from the tropical storm in the Atlantic. Landfall is expected to occur just north of your town in three days. A meeting has been called so that operations personnel can determine what, if anything, needs to be done to secure the station and prepare for the expected increased call load.

"Are you available to staff a unit starting the night before expected landfall?" your captain asks you.

"I don't see why not. I just have to let my wife know," you respond.

You cannot help but think to yourself that this prep work is all for nothing. The last time a storm hit your town it ended up dropping a much smaller amount of rain then forecast and a bunch of people sat around staring at each other, waiting for an emergency. But, if the leadership wants to get ready, you will be there.

Why do agency leaders start planning so far in advance for a known event? Are there really that many changes to response activities during a major event versus daily response activities? What happens if this event does get too big for your agency, or even jurisdiction, to respond properly? What resources are there to help?

Introduction

We are familiar with the news coverage and pictures of the response to major disasters across the country. From natural to manmade disasters, the potential need exists across the country to respond to such a situation. The question is, are you ready for a disaster? It seems like such a simple question, but when you consider your personal preparedness as well as your knowledge and skill for professional response, it may not be so easy to answer. Disaster response is a physical and emotional, time-consuming activity that can take you away from home for days, weeks, or months on end. It requires the response of not only local agencies, but also regional, state, and federal entities to assist in the response and recovery stages. Disasters mean that you will work with people you have just met and face significant issues that you may not have trained for. However, it is your responsibility to continue providing quality care during these incidents, as well as fill roles that you might not find standard to your job description. Understanding the organization and the process of disaster response is important in ensuring the safety of the citizens of your locality.

When a significant event such as a hurricane, tornado, or wildfire occurs, it is easy to refer to the incident as a disaster. Describing incidents that go above and beyond a normal emergency or a mass casualty, we tend to use the word "disaster." While you may deem many significant incidents within your jurisdiction disasters, the term "disaster" has a legal definition. The Robert T. Stafford Act (2007) defines a major disaster as "any natural catastrophe . . . , or, regardless of cause, any fire, flood, or explosion, in any part of the United States, which in the determination of the President causes damage of sufficient severity and magnitude to warrant major disaster assistance under this Act to supplement the efforts and available resources of States, local governments, and disaster relief organizations in alleviating the damage, loss, hardship, or suffering caused thereby."[1] By this definition, an event can only receive a disaster declaration from the President. A specific process, as set forth in the Stafford Act, must be followed in order for an incident to receive a disaster declaration and all the rights that declaration provides, including money and resources to assist with response, recovery, and mitigation. In your daily incidents, mutual aid is not always a necessity. In a disaster, you are almost guaranteed the need for assistance from assets outside your locality, whether fire, EMS, police, public works, or other agencies. When your agency relays the need for assistance through the appropriate channels, the requests are made to get the necessary resources.

Mutual Aid Assistance

The first level of assistance is local mutual aid. Mutual aid provides units from surrounding agencies and jurisdictions to assist in incident response. There are three types of mutual aid utilized to obtain additional resources: automatic, regional, and statewide.

Automatic aid

Automatic mutual aid is set up prior to use and normally involves both jurisdictions signing an agreement that covers aspects of liability, reimbursement, and expectations. It involves the automatic response of neighboring jurisdictions based on earlier agreements. Incidents that may require the use of automatic aid include house fires and interstate incidents. Automatic mutual aid is normally set up during these types of incidents so that the closest apparatus can respond, regardless of jurisdiction. Automatic mutual aid is often requested when the incident is still small scale and has not required the activation of the local or state emergency operations center (EOC). Automatic mutual aid can, however, be requested after EOC activation. If this occurs, it should still be requested through standard dispatch center policies.

Regional mutual aid

Regional mutual aid is the second type of mutual aid that can be utilized in incident response. Regional mutual aid turns to jurisdictions beyond those neighboring your locality. Regional mutual aid is often used in an event that is of long duration but a small area of impact. The resources generally come from a variety of agencies within an area or from a business or agency that serves a regional area. A regional hazardous material team is an example of a mutual aid resource. A hazardous materials team is often comprised of representatives from multiple area agencies. This allows for cooperative response between agencies and also ensures that one agency is not overwhelmed during response to an incident. Resources obtained through regional mutual aid are typically dispatched through the local public safety answering point (PSAP). However, if regional mutual aid is determined to be necessary during a declared state of emergency, then the local EOC may handle the mutual aid request process.

Statewide mutual aid

Statewide mutual aid is the third type of mutual aid available to assist localities in incident response when local and regional mutual aid resources are not available or adequate. The definition of statewide mutual aid differs from state to state, but the basic idea is the same. Statewide mutual aid may involve the request of resources from across the state or the use of state sponsored assets to respond to the incident. State assets may be ambulances or personnel who are trained and prepared for specific types of incidents. When statewide assistance involves the use of local assets, it is normally accomplished through an agreement between participating localities, similar to that used during automatic mutual aid. The agreement covers issues such as liability insurance, reimbursement, appropriate use of resources, and communications. These agreements ensure that any "red tape" issues are handled prior to the deployment of resources. The difference in automatic aid and statewide mutual aid is that when requesting statewide mutual aid assistance, resources may come from any part of the state, not just the immediate response area.

Emergency Management Assistance Compact

The Emergency Management Assistance Compact (EMAC) is an organization that provides another level of mutual aid assistance during large-scale/disaster situations. EMAC was established in 1996 to provide a structure of deployment of assistance for all its member U.S. states and territories.[2] There are five distinct phases to the EMAC process, each of which provides a step to ensuring quick and efficient assistance for states in need. These phases range from pre-event preparation and planning to reimbursement for the cost of deploying resources. Each phase supports the EMAC structure to provide effective and efficient resources to requesting states.

Emergency Operations Centers

A small-scale incident can be handled through the use of standard policies and procedures of a locality. The facilities, personnel, and resources that are available on a daily basis are utilized to mitigate the ongoing situation. On-site, the incident command structure is utilized to manage the single site operations and an incident command post serves as the central point of information gathering and dissemination and as the central gathering point for decision makers. Because a large-scale incident may expand to multiple locations, multiple ICS structures may develop to assist in response. Even if multiple ICS structures are not created (the incident occurs in one location), the scale of the incident dictates the need for coordination beyond what is provided by the incident command post. Area commands are normally established when multiple ICS structures are developed to handle an incident. As previously discussed, these area commands must coordinate with each other to ensure that response is effective and efficient. When an incident grows to this size, additional facilities, personnel, and resources are necessary to successfully coordinate incident response and recovery efforts. When the need arises, affected localities activate an emergency operations center (EOC) to provide coordination of personnel and resources on a large scale.

The EOC is an integral part of the coordinated effort of response during disaster situations. Your locality's EOC provides a central location for all representatives of government to gather and coordinate response and recovery efforts within the jurisdiction.[3] EOC operations allow for the coordinated efforts of public and private entities without requiring those people to respond to the scene and provide assistance from the ICP. An EOC may be activated by a single agency, a locality, or the state government; each providing a site for coordination of resource request and response activities.

EOC purpose

While it seems obvious, the purpose of the EOC is important to remember in its development, activation, use, and demobilization. Each step of the response process is important, but remembering why you need an emergency operations center is important in making sure you create a beneficial facility to support incident response activities. When taking a general approach to establishing an EOC, remember that the purpose of an emergency operations center is "to establish a central location where government at any level can provide interagency coordination and execute decision making to support incident response."[4] A well-established and utilized EOC ensures that the appropriate people are gathered together to receive, analyze, and interpret information to ensure that decisions impacting incident response are appropriate, effective, and efficient in mitigating an emergency situation. These are decisions that, when performed on-site, may be limited in effectiveness or affected and influenced by current on-scene operations.

On a smaller scale, the EOC allows people to work while maintaining a "big picture." While on-scene, individuals are often distracted by the sights, sounds, and activities of other responders. Placing the decision makers in a facility that provides them with real-time information while separating them from ongoing activities helps to ensure that they can work effectively and efficiently.

EOC activation

Not every incident will result in activation of the emergency operations center, and not every activation of the EOC involves EMS personnel. EOC activation occurs at the discretion of local leadership and emergency management personnel. Determination of activation trigger points is based on predetermined policies and procedures, but they should be specific, measurable, realistic, and time sensitive, allowing for some interpretation but maintaining a specific nature.

JUSTIFICATION FOR OPENING THE EOC

1. A unified command or area command is established.

2. More than one jurisdiction is involved in the response to an incident.

3. An incident commander recognizes that an event requires significant resources and assistance and could escalate rapidly.

4. Historical data shows that similar events were mitigated with EOC activation.

5. Appropriate leadership directs for the activation of the EOC.

6. A known emergency event occurs (e.g., hurricane, flooding, snow storm, etc.).

7. The local emergency operations plan (EOP) directs EOC activation.

If an event is determined to be imminent, such as an incoming hurricane, the emergency operations center may be activated early, allowing for the coordination of personnel, resources, and equipment prior to the actual event. If an event occurs without warning, such as a terrorist incident, the EOC will be activated after the event. This can lead to a delay in the coordination of resource requests for incident response. While you may not be responsible for the activation of the emergency operations center, it is beneficial to know when to expect EOC activation, as you may find yourself with new responsibilities both on-scene and at the EOC when it is activated.

EOC staffing

Once the decision to open the EOC is made, the determination of staffing is the next decision. The first issue is which functions must be staffed within the EOC. Just because your jurisdiction opens the EOC does not mean that EMS representation is necessary. Staffing of the medical/EMS chair is necessary when there is an expected health and medical impact during an incident. When staffed, the EMS/medical representative serves as a liaison for the EMS response staff in the field to initiate resource requests.

You may wonder how the staffing of the health and medical position in the EOC impacts you. First, if you hold a leadership position within your EMS agency, you may become the individual requested to staff the position or the individual responsible for delegating authority to someone to serve in the EOC. If you fill an operational position (work in the field) with your agency, you may be speaking directly with the EOC staff during a large-scale incident. Knowing who is serving in the health and medical position and how they can assist you is important in ensuring continuous response during an incident.

The individuals typically chosen to fill staff positions in an EOC (local or state) have certain characteristics that make them successful. Regardless of the purpose of the phone call, you should consider the following characteristics necessary for EOC personnel. The individuals chosen to staff the health and medical position, or for that matter any position in the EOC, should possess the knowledge, skills, and abilities necessary to effectively serve in the position.[5] Training requirements should be predetermined prior to the need for staffing. Requirements for training should include an understanding of basic EOC management, an understanding of the specific aspects of the ESF-8/EMS chair, an understanding of the ICS structure, and an understanding of the resource request process, including FEMA requirements. Basic training can be acquired by many methods, including through the National Fire Academy (NFA), the Emergency Management Institute (EMI), and the state Office of Emergency Management. Individuals selected to serve in the EOC should also have authority to perform the necessary tasks. If you do not have authorization to request mutual aid EMS resources, then your assignment should not be in the EOC. However, if you are chosen to staff the EOC, you may be delegated the authority to make decisions that support response activities.

Staffing numbers is also a consideration when writing policies and procedures for EOC activation. It is feasible to consider that in a smaller locality, a single person may be utilized to represent multiple areas of function within the local EOC. For example, a lieutenant within a fire and EMS company may serve in both the search and rescue and the fire department position. If you are chosen to serve in the EOC, you may represent all of the emergency services functions, not just the EMS function. Staffing will be dependent on the size of the incident and the needs that have arisen as a result. While FEMA makes recommendations for full-time staffing of the emergency operations functions within a locality (table 4–1), usually only one or two individuals are assigned to staff the emergency management positions on a daily basis. This will obviously change as the need for personnel and EOC support increases within the locality.

Table 4–1 Recommended full-time staffing levels for emergency operations center

Locality Population	Full-Time Staff
Over 1 million	6–20
250,000	4–8
100,000 to 250,000	3–5
25,000 to 100,000	2–3
Under 25,000	1–2

One way to ensure that appropriate staff are selected to serve in the various positions of the EOC is to create position descriptions. A well-written position description will provide a list of general responsibilities and begins the framework of training requirements for any personnel staffing the position. A position description should be written for all positions in the EOC, not just the medical/EMS position.

Organizing the EOC

Whether you serve in the EOC or interact with it while conducting operations on-scene, understanding the organization of the EOC personnel and the communications flow to the appropriate individuals as well as within the EOC is important. There are many different layouts and organizational methods to EOC management. If you interact with multiple EOCs you may see that each has a different structure. Being familiar with their structure and policies is important in emergency response.

Organization by ICS standards. One option for EOC structure is to organize staffing based on the ICS standards and recommendations. In this structure, the EOC staff mirrors the on-site personnel structure through the incident command system. The leader of the EOC holds a similar role to the incident commander in the field; however, instead of oversight of field responders, the EOC leadership provides oversight to the policy and decision makers. Similarly, organization within the EOC should mirror the field ICS structure. For instance, the grouping of the emergency support functions should be the same as the ICS structure.

1. Operations. The representatives in this function focus on coordinating and supporting the on-scene responders. The individuals assigned to this function may secure additional resources through mutual aid agreements. As a member of ESF-8, you would likely be assigned to this group.

2. Planning. As with the planning function at the incident scene, representatives in the planning section at the EOC focus on gathering, analyzing, and disseminating information associated with the incident. In an event such as a hurricane, incidents may be widespread throughout your locality. The planning section can maintain information on all of the incidents, ensuring the local leadership is aware of all response activities.

3. Logistics. Members serving under the logistics function provide support to both personnel in the EOC and those working in the field. The logistics function in the EOC focuses on securing communications within the locality, lodging, transportation, and other necessities.

4. Finance/administration. The finance/administration function coordinates all financial management activities associated with the locality's response to the incident. If the incident receives a federal emergency declaration, reimbursement of expenses through FEMA is dependent on appropriate tracking of time and spending.

The major benefit of this organizational method is that it provides a direct communication line from resources in the field to resources at the local EOC. There is a known system of communication and who field personnel should communicate with. Resource requests may run in a more seamless manner when utilizing this organizational structure. While the ICS method for organizing the EOC best fits with interaction of field resources, there are alternate structures that localities utilize to organize the EOC.

Organization by major activity. Many representatives within the EOC conduct activities that fall within a variety of categories. One method for organizing the EOC is to group those representatives whose major activities are similar, so that they can interact and coordinate response activities to ensure that duplication of effort does not occur. In this organizational method, the following categories are used:

- Policy group. Comprised of the key leadership of a jurisdiction, this group makes decisions regarding policies and procedures of the entire incident, not just a small part of the emergency.

- Resource group. This group is comprised of those agencies that have equipment, personnel, and other resources that may assist in response activities. Remember that private organizations with resources should also be represented in the EOC in this group.

- Operations group. The individuals in this group are those who have responsibilities in the response activities during the incident. Agencies in this group are often represented both in the operations group and the resource group. Agencies involved in this group may include, but are not limited to, fire, EMS, police, and public works. The scope of the incident will dictate which agencies are represented in the operations group.

- Coordination group. This group is responsible for the data analysis and damage assessment of the areas impacted by the incident. The group may also provide predictions for future needs and issues that other EOC groups can then use for planning.

While this system provides a simple method for organization and ensures that the key individuals are represented within the EOC, there is a major disadvantage. The interaction between on-site ICS command personnel and the appropriate EOC area may be confusing. The structure does not allow for a simple communication plan from on-site operations to EOC operations. This type of communication system would require additional planning that may change from incident to incident.

Organization by ESF. ESF stands for emergency support function. At the federal level, resource management is organized through defined emergency support functions. These functions vary from medical to logistics to damage assessments and everything in between. These functions are described in depth later in this chapter. Because of the expectation of interaction with federal resources during a large-scale emergency event, many states have taken steps to organize the state EOC based on the emergency support functions of the representatives assigned to the EOC. When possible, many local EOCs use this organizational method as well to provide a seamless line of communication from the local EOC to the state EOC and on to federal resources. While FEMA and other federal response agencies recognize 15 emergency support functions, they are combined into groups that mirror the ICS structure.

1. Operations. The operations group includes the following functions/units:

   a. Public works/emergency engineering branch

   b. Firefighting branch

   c. Public health and medical services branch

   d. Urban search and rescue branch

   e. Public safety/law enforcement branch

2. Planning. This group includes the following functions/units:

   a. Situation analysis unit

   b. Documentation unit

   c. Advanced planning unit

   d. Technical services unit

   e. Damage assessment unit

   f. Resource status unit

   g. Geographic information systems (GIS) unit

3. Logistics. While not necessarily tied directly to an ESF staff member, the logistics functions are staffed by representatives of multiple ESFs:

    a. Situation analysis unit

    b. Communications unit

    c. Food unit

    d. Medical unit

    e. Transportation unit

    f. Supply unit

    g. Facilities unit

4. Finance/administrative. The finance/administrative group includes the following functions/units:

    a. Compensation claims unit

    b. Cost unit

    c. Purchasing/procurement unit

    d. Time unit

    e. Disaster financial assistance

While this structure provides guidance so on-scene ICS personnel understand who they interact with in the local EOC, and the local EOC representatives understand who they interact with at a state level, there are some disadvantages. The first is that while states may use ESFs for resource organization, their ESF groupings do not match those at the federal level, leading to some confusion. A second disadvantage is related more directly to the training and preparedness activities associated with EOC activation. In order to ensure that individuals tasked with staffing an ESF function at the EOC are best prepared, intensive training is required. Localities will then have to train a large number of people, incurring a large cost, or have only a small pool of personnel to pull from for EOC staffing.

Organization by multiagency coordination system (MACS). A multiagency coordination system (MACS) is a combination of agencies, organization, people, and resources combined for a single response to an incident.[6] In organizing an EOC through a MACS, the staffing is determined by need, not necessarily by the organization of the functions. In some incidents, it is easy to determine the appropriate representatives, while other incidents require a bit more consideration of staffing representation. The huge disadvantage to organizing as a MACS is that there is no clear definition of expectations and roles within the EOC. While a MACS coordinator may be selected to provide guidance, it is only a recommended role, not a required one. There is no clear relationship between the MACS and the on-site ICS structures, which leads to limited and confusing lines of communication.

EOC relationship with ICS structure

The EOC serves as a critical link in response to emergency incidents. It allows for a well-organized management of deployed resources and provides a central point for multiagency coordination efforts. Jurisdictions with well established EOC procedures and guidelines have found that they have a more efficient communication line from on-scene to EOC operations, and that their on-scene incident commander can focus on the operations at the incident instead of the logistics of resource requests and provide strategic guidance in support of incident operations. Through the development of NIMS procedures and protocols, emergency operations centers become an integrated part of the incident command structure and assist in multiagency coordination.

While an EOC is established, the expectation is that interaction with on-scene resources meets and/or exceeds the expectations set forth in NIMS documents. The EOC must integrate its activities smoothly while ensuring the ICS needs at the incident site are met. Activities of the EOC should support the following needs:

- Common operating picture. At an incident, every agency is focused on the same final goal, it is just the steps to get there that are different. In a large and/or complex situation, the coordination of a large number of agencies is best managed off-site by the operation of an EOC.

- Policy direction. When you have brought in multiple agencies and multiple jurisdictions to respond to an incident, it is inevitable that a conflict of policies will occur. On-site policy direction would take the focus away from direct operational aspects of incident response. Because of this, an EOC allows for jurisdictions and agencies to meet and reach compromise on policies without interfering with on-scene incident operations.

- Communication support. The network necessary to support a large-scale operation includes not only local public safety answering points (PSAPs), but also an integrated communication system between the locality where the incident occurs and the agencies supporting the localities. PSAPs require logistical support to ensure that all incoming resources can communicate once on-site and that situation updates are communicated through the appropriate channels.

- Resources. The EOC integrates with the incident ICS structure by providing the communication with the localities and agencies to secure appropriate resources. As will be discussed in further detail later in this chapter, upon activation of an EOC, resource requests are funneled through the EOC to ensure proper tracking, management, and procedures are followed and to allow on-site personnel to focus on continued incident response.

- Strategic planning. Long-term planning is likely necessary once a large-scale incident occurs. Planning is best accomplished away from the incident scene. Incident operational planning occurs at the incident command post, with on-site management personnel participating. However, long-term strategic planning may require additional resources and personnel, including county administration, state agencies, and other personnel, to ensure adequate plans are completed.

- Legal and financial support. A separate branch of the incident command structure provides support in tracking and completion of financial paperwork. During a small incident these activities are often completed off-site. During a large incident, the EOC provides a central location for personnel assigned to multiple financial/administration branches to ensure appropriate completion of paperwork on a larger scale. It also ensures a quicker exchange of information necessary to complete some paperwork.

With the full integration of an EOC into incident operations, it becomes part of the multiagency coordination system (MACS). As previously mentioned, a MACS is not a facility but an organization of response resources and personnel. As a part of a MACS, EOC staff are responsible for ensuring, when necessary, coordination between other MACS entities occurs. Other MAC entities may include EOCs in other localities, the state EOC, and area command posts from the various incident sites. Coordination of activities may include mutual aid requests (ensuring that EOCs from multiple jurisdictions are working together in their requests), and deployment and involvement of technical specialists (including but not limited to those provided by state environmental agencies for hazmat situations, outside contractors, or researchers). Another situation that requires the coordination of EOC with other MACS entities is a widespread or federally declared disaster. These situations normally involve the immediate activation of MACS elements and entities, forcing the need for cooperation between EOCs and incident operations.

Stress and Emergency Response

Stress can have a negative impact on an individual's ability to make sound decisions, especially in a split second. The stress associated with response to emergency situations, especially large and complex situations involving activation of the EOC, can negatively impact your ability to complete your objectives. Whether serving in the EOC or participating in incident response operations, it is important to understand and recognize situations in which stress is influencing your ability to work. Remember too that stress levels will not only impact you personally, but also your ability to work and interact with others within your workspace (field or EOC). Stress can lead to personal conflict, sensory overload (and therefore lack of attention), and perception distortion coupled with poor judgment. Developing methods to recognize and minimize stress is important in maintaining safe and effective operations.

One of the first steps in managing stress is to make sure you can recognize the signs of stress in yourself and other emergency services personnel working at the scene or in the EOC. Individuals who are experiencing negative stress effects display similar characteristics. One characteristic is that the individual makes rash decisions.[7] Because of high stress levels, the individual may not wait for the appropriate information and situation reports on which to base his or her decisions. A second characteristic pointing to stress effects is that the person's reactions tend toward aggression. Instead of being able to cope with stressful situations, the individual may react with rage, anger, or aggressive responses instead of a standard response (for that individual).

Another concerning characteristic is that an individual who has normally been able to handle difficult tasks is no longer able to do so. While he or she may handle simple tasks such as filing and computer entry, requesting participation in data analysis or development of situation reports may result in an incomplete or inadequately completed task.

Creating methods for minimizing the effects of stress is important, as is putting into place methods for providing assistance to those who have been negatively impacted by the incident. The first step is to ensure that staffing levels allow on-shift personnel to be able to take breaks from the ongoing activities. Constant exposure to the stressful situations caused by an emergency incident without breaks can increase the stress an individual feels. It is also important to continue to monitor those who participated in incident response, from the PSAP workers to the personnel staffing the EOC. Stress effects may be immediate, but often the effects will not be seen until later. Monitoring those responders using both self-evaluation and the evaluations of individuals by their managers and coworkers can provide early recognition of those who are experiencing negative stress effects. Once it is recognized someone is negatively impacted by the incident, offer help. Help may come in the form of critical incident stress management (CISM) techniques or through the offering of professional mental health services. Regardless, once an individual has been recognized as being negatively impacted by the stressful events of the incident, providing assistance in dealing with the impact is imperative.

Requesting EMS Resources

Scenario: After attempting to get some sleep while the storm raged outside the EOC, you return for your next shift. Upon your arrival you are briefed by the off-shifting EMS representative.

"Calls have been pouring in. Residents are calling because they chose not to evacuate, have had second thoughts, and now cannot get out due to the rapidly rising water. We have received updates from all of our stations. Two have experienced flooding within the buildings, three ambulances have been flooded, and one fire truck is a loss due to a tree falling on it. Thankfully no providers have been injured in these incidents. We also heard from the hospital that they relying on their generators for electricity but have concerns about how long they will last with the fuel they have."

After finishing the briefing, you sit down to assess the situation. Everything seems a bit overwhelming, but not unmanageable. Making sure that you are taking on assignments that are appropriate to your role is important, but the only place to start is from the beginning. As you sit down to make a to-do list, EMS-1, the on-duty EMS supervisor, calls you.

"We have a problem. Our people have not been home in two days and they need a break. We need to find resources to cover our EMS calls while our people go home and assess the damage to their homes and impact of the storm on their families."

What resources are available to assist your crews? Where can you go to get assistance and provide an opportunity for your staff to get a break, but also still provide coverage for the citizens of your locality?

It seems simple. When there is a need for help during standard day-to-day operations, a call goes out to the ambulance at the next closest station. A disaster, whether natural or man-made, is not a standard day-to-day operation. Not only are working conditions significantly different from those you are used to, but the resources that you would normally call on are unable to assist because they too have been impacted or have already been utilized and are unavailable due to the duration of the event. Many people believe that once assistance is needed the federal government will step in to provide resources. This is not the case, and having knowledge of the request process can assist you in structuring your response and requesting the appropriate resources at the appropriate time.

Local medical resources

While an EMS provider has access to many resources through standard mutual aid processes, knowing the potential medical resources available to assist in response to an emergency is critical. This will not only allow you to understand what is available, but also provide the ability to make a more specific request when help is necessary. Knowledge of the medical assets available to your agency also allows you to meet and develop relationships with these assets prior to using them in the field.

Community emergency response team (CERT). In 1985, after returning from a trip to Kobe, Japan, and participating in response to an earthquake in Mexico, the Los Angeles Fire Department (LAFD) began developing a program that provided training to community members to assist in response to emergency events.[8] During their trips, members of the LAFD saw two very different types of response. In Japan, members of tight-knit communities had trained and developed plans for response to emergency incidents within their community. In Mexico, while the response of the members of the community was successful in saving a large number of victims, training was minimal and organization of response was nonexistent. With this in mind, the original CERT program developed by the LAFD was designed to provide an understanding to citizens of their role during a disaster, specifically earthquakes, and assist in providing education on family, friend, and personal preparedness. In 2002, the push for local community response involvement came from President Bush. During a speech post 9-11, President Bush challenged Americans to become involved in their communities.

With the help of FEMA and the Emergency Management Institute (EMI), the CERT training took on an all-hazards approach to preparedness. CERT members are trained in basic disaster preparedness, fire suppression, disaster medical operations, light search and rescue operations, and disaster psychology.[9] At the completion of the program, students participate in a mock disaster where they put the skills taught in class into practice. CERT members are trained in NIMS and ICS and can provide assistance during all phases of incident response. In a large-scale disaster such as a hurricane, the local CERT members may be unavailable to assist. While they may not be able to provide in-depth medical assistance, CERT members can provide basic staffing. If necessary, area CERT members may be requested to assist in filling positions such as shelter oversight and search and rescue support. Because medical support resources may be limited due to the significant impact of an event, any available and trained resource should be considered to maximize response efforts.

According to www.citizencorps.gov, there are currently 3,506 registered and active community emergency response teams in the United States.

Metropolitan medical response system (MMRS). The metropolitan medical response system (MMRS) is an operational system developed to respond to terrorist incidents and other public health emergencies that create mass casualties or casualties requiring unique care capabilities.[10] Currently funded by the U.S. Department of Homeland Security grant program, MMRS teams are designed to provide mass casualty incident response management prior to the arrival of state and federal assistance. The MMRS program is a locally based system often developed through the partnership of neighboring jurisdictions. The pilot concept was developed in both Washington, DC, and Atlanta, Georgia, from 1995–1996, before Congress granted permission for the Department of Health and Human Services (DHHS) to develop 124 programs across the country.

The resources of a MMRS program are vast. Not designed to take over the response capabilities of any locality, the MMRS provides an opportunity to include in the planning and response phase partners who might not have previously been considered, such as private doctors and medical care providers. The resources provided through federal grants can assist localities in response to any and all hazards including hazardous materials, building collapse, and other significant incidents. You should meet with the local MMRS program to determine the availability of response and the assistance that can be provided. Under grant requirements, MMRS programs must work with the local hospital system to ensure appropriate equipment, medical staff, and resources are available to assist in response. They are tasked with ensuring a process for forward movement of patients, which is normally done through established local processes. The benefit of the MMRS is that they can provide the resources to create a field hospital and assist in triage, treatment, and transport of patients from the incident scene.

As of July 2011 there are a large number of MRC units registered across the United States. The current number of teams per FEMA region are as follows:

- Region I – 85

- Region II – 58

- Region III – 79

- Region IV – 154

- Region V – 240

- Region VI – 97

- Region VII – 65

- Region VIII – 45

- Region IX – 61

- Region X – 51

Medical Reserve Corps (MRC). Sponsored by the office of the surgeon general, the Medical Reserve Corps (MRC) is designed to coordinate the training and response of medical personnel who respond during an emergency as a supplement to public health resources.[11] These volunteers are trained personnel from all different fields including pharmacists, pediatricians, cardiologists, general practitioners, dentists, veterinarians, and BLS and ALS providers. When there is no active emergency response necessary, MRC teams provide public health and wellness education and training to members of the community.

MRC members are a great medical asset. Not only are members trained in treatment of patients, some with significant specialty training, but they are also trained and prepared to assist in various aspects of emergency response outside of patient care, including shelter management and mass immunization. They can also provide assistance that may be necessary beyond direct patient care. Some teams have chaplains that can provide counseling and mental health services, lawyers who can advise on protocol and standard of care concerns, office workers who can assist in record keeping and documentation, and even interpreters who can assist in crossing language barriers. Each of these resources may be necessary to ensure that your incident is effectively and efficiently controlled.

Federal Medical Resources

Scenario: It has been 48 hours since hurricane Daniel made landfall just 50 miles north of your town. Your agency has continued to run the calls, and with the help of state assistance have weathered any problems to this point. People are tired, though, and you are dreaming about your downtime, when the phone at your desk in the EOC rings.

"EOC, this is Lieutenant Johnson. What can I do for you?"

"Lieutenant Johnson, my name is Jim Smith and I am the operations manager for North Star Hospital. We are facing a major issue and need some resources to help us out."

"What seems to be the problem?" you ask while crossing your fingers that it will be a simple fix.

"Well, we have one generator that has failed, another that is about to fail, and a third that is low on gas. We are preparing to activate our emergency evacuation plan, but are coming up short on resources. We are in need of assistance of moving these patients, and even in finding a place to move them to that isn't already impacted."

After gathering all of the necessary details, you approach your supervisor and begin determining the best plan of action. Knowing that your local and regional resources are currently overwhelmed or on down time from working long hours, you begin to discuss the other resources that may be available to assist North Star Hospital.

With all the resources you know currently involved in response activities, what is left? Who do you call for a request of this nature and this size? What can you do to make sure that the patients at North Star Hospital are taken care of?

Once the governor of a state has determined that the statewide resources are no longer capable of responding to the disaster effectively, a request to declare the incident a disaster is sent to FEMA and the President for consideration. The request and ensuing declaration is based on the damage assessment, and contingent on the governor's commitment of continued use of state funds and resources in the response and long-term recovery from the incident (http://www.fema.gov/hazard/dproc.shtm).[12] Once the commitment of resources is made, response of federal assets is coordinated through the use of groupings based on the functions that are being conducted on the scene.

When large incidents that have a potential to require federal assistance (whether EMS or other) occur, the National Response Coordination Center (NRCC) serves as the coordination point for all federal requests and activation of lead federal agencies. The NRCC is a small component of the National Operations Center (NOC) and serves as the Department of Homeland Security/Federal Emergency Management Agency primary operations center responsible for national incident response and recovery as well as national resource coordination.[13] Similar to a local/state EOC, the NRCC serves as the central point for federal resource coordination during incidents that receive national disaster declarations. Because of the large number of federal agencies potentially involved in incident response, federal resources are organized based on the function and activities that they participate in during response activities. These categories are known as emergency support functions (ESF).

Emergency support function structure. In order to provide some guidance and structure in the deployment and response of federal assets, federal agencies and organizations are categorized based on the activities that they support and are involved in during an incident. At the federal level, there are 15 ESF categories, each with a different scope of work (table 4–2). As previously mentioned, states will often mirror their EOCs and state agency organization to match the ESF structure at the federal level to provide a smooth communication system between the state EOC and federal agencies during incident response. Each ESF also has a coordinating agency that serves as the organizing force for all activities within that support function (table 4–3).

Table 4–2 Federal emergency support functions scope

Emergency Support Function	Scope
ESF #1 – Transportation	Aviation/airspace management and control Transportation safety Restoration/recovery of transportation infrastructure Movement restrictions Damage and impact assessment
ESF #2 – Communications	Coordination with telecommunications and information technology industries Restoration and repair of telecommunications infrastructure Protection, restoration, and sustainment of national cyber and information technology resources Oversight of communications within the federal incident management and response structures
ESF #3 – Public Works and Engineering	Infrastructure protection and emergency repair Infrastructure restoration Engineering services and construction management Emergency contracting support for life-saving and life-sustaining services
ESF #4 – Firefighting	Coordination of federal firefighting activities Support to wildland, rural, and urban firefighting operations
ESF #5 – Emergency Management	Coordination of incident management and response efforts Issuance of mission assignments Resource and human capital Incident action planning Financial management

Emergency Support Function	Scope
ESF #6 – Mass Care, Emergency Assistance, Housing, and Human Services	Mass care Emergency Assistance Disaster housing Human services
ESF #7 – Logistics Management and Resource Support	Comprehensive, national incident logistics planning, management, and sustainment capability Resource support (facility space, office equipment and supplies, contracting services, etc.)
ESF #8 – Public Health and Medical Services	Public health Medical Mental health services Mass fatality management
ESF #9 – Search and Rescue	Life-saving assistance Search and rescue operations
ESF #10 – Oil and Hazardous Materials Response	Oil and hazardous materials (chemical, biological, radiological, etc.) response Environmental short- and long-term cleanup
ESF #11 – Agriculture and Natural Resources	Nutrition assistance Animal and plant disease and pest response Food safety and security Natural and cultural resources and historic properties protection and restoration Safety and well-being of household pets
ESF #12 – Energy	Energy infrastructure assessment, repair, and restoration Energy industry utilities coordination Energy forecast
ESF #13 – Public Safety and Security	Facility and resource security Security planning and technical resource assistance Public safety and security support Support to access, traffic, and crowd control
ESF #14 –Long-Term Community Recovery	Social and economic community impact assessment Long-term community recovery assistance to states, local governments, and the private sector Analysis and review of mitigation program implementation
ESF #15 – External Affairs	Emergency public information and protective action guidance Media and community relations Congressional and international affairs Tribal and insular affairs

http://www.fema.gov/pdf/emergency/nrf/nrf-esf-intro.pdf

Table 4–3 Coordinating agency for each ESF

ESF Number	Function	Coordinating Agency or Agencies
1	Transportation	Department of Transportation
2	Communications	Department of Homeland Security National Communications System Federal Emergency Management Agency
3	Public Works and Engineering	Department of Defense U.S. Army Corps of Engineers
4	Firefighting	U.S. Department of Agriculture Forestry Service
5	Emergency Management	Federal Emergency Management Agency
6	Mass Care, Emergency Assistance, Housing, and Human Services	Federal Emergency Management Agency
7	Logistics Management and Resource Support	Federal Emergency Management Agency U.S. General Services Administration
8	Public Health and Medical Services	Health and Human Services
9	Search and Rescue	Federal Emergency Management Agency
10	Oil and Hazardous Materials Response	Environmental Protection Agency
11	Agriculture and Natural Resources	U.S. Department of Agriculture
12	Energy	Department of Energy
13	Public Safety and Security	Department of Justice
14	Long-Term Community Recovery	Federal Emergency Management Agency
15	External Affairs	Department of Homeland Security

The coordination agency serves as the contact point and primary distributor of information to agencies that support the response activities within that specific function. The coordinating agency provides management oversight for their ESF through the preparedness, response, and recovery phases of an incident. The coordinator serves in the "unified command" structure and provides feedback up the chain to assist in coordination of federal asset response.[14] Not only does the ESF coordinating agency provide oversight to their specific ESF, but they also ensure that agencies within the ESF are communicating with each other and coordinating response activities to ensure there is no duplication of response efforts.[15]

The ESF structure at the federal level also designates primary agencies in response efforts. The primary agency is a federal, not a private, agency that holds "significant authorities, roles, resources, or capabilities for a particular function."[16] These agencies are able to provide the primary support for the state requests received within their ESF. Primary agencies work to provide assistance to a requesting state within the emergency support function they represent. They coordinate federal response efforts with state assets and ensure that the appropriate resources have been dispatched. Often a primary agency also serves as the ESF coordinator. Table 4–4 provides a list of the primary agencies for each ESF. Along with the primary and coordinating agency, support agencies ensure that the work that needs to be completed and the requests that are made are handled in an efficient and timely manner so that the locality can recover from the incident.

Table 4–4 Primary agency for each ESF

ESF Number	Primary Agency
1	Department of Transportation
2	Department of Homeland Security National Communications System Federal Emergency Management Agency
3	Department of Defense U.S. Army Corps of Engineers
4	U.S. Department of Agriculture/Forestry Service
5	Federal Emergency Management Agency
6	Federal Emergency Management Agency
7	Federal Emergency Management Agency U.S. General Services Administration
8	Health and Human Services
9	Federal Emergency Management Agency Department of Defense
10	Environmental Protection Agency
11	U.S. Department of Agriculture/Department of Interior
12	Department of Energy
13	Department of Justice
14	U.S. Department of Agriculture Department of Homeland Security Housing and Urban Development Small Business Association
15	Federal Emergency Management Agency

ESF-8. Deployment of federal resources to assist with the health and medical needs of an impacted area is coordinated through ESF-8, Public Health and Medical Services (figure 4–1).

Fig. 4–1 ESF-8 can request resources to assist in providing medical assistance during a disaster.

With the Department of Health and Human Services (DHHS) acting as the coordinating and primary response agency, ESF-8 is response for the following during disaster operations:

- Assessment of public health/medical needs

- Health surveillance

- Medical care personnel

- Health/medical/veterinary equipment and supplies

- Patient evacuation

- Patient care

- Safety and security of drugs, biologics, and medical devices

- Blood and blood products

- Food safety and security

- Agriculture safety and security

- All-hazard public health and medical consultation, technical assistance, and support

- Behavioral health care

- Public health and medical information

- Vector control

- Potable water/wastewater and solid waste disposal

- Mass fatality management, victim identification, and decontaminating remains

- Veterinary medical support

The federal agencies supporting ESF-8 activities respond using protocols set forth by the secretary of Health and Human Services (HHS). These protocols allow for the deployment and actions of assets in the field. While the secretary of Health and Human Services serves as the leader of ESF-8 response, the assistant secretary for Preparedness and Response serves as the coordinator of ESF-8 agencies and response activities.[17] Staffing is immediately augmented at the notification of a possible or actual event so that HHS can begin monitoring the situation and determining public health and medical services needs within the affected locality.

While public health and medical services needs go beyond direct patient care, the focus of the EMS is on the patient care, patient evacuation, and mass fatality management, as those are the operational aspects that most fit the activities you will be carrying out during a disaster. Through the request process, your state may determine there is a need for additional medical resources (from EMS providers to physicians) and as such may request one of the DHHS assets that are available in disaster situations. As an EMS provider, you must understand the capabilities and functions of these assets so that you can work to integrate them into the ICS structure that has been formed.

Areas of Medical Function Specific to EMS

- **Medical care personnel.** ESF-8 can request internal HHS assets (National Disaster Medical System, federal civil service employees) and personnel from other assisting agencies to provide medical assistance. Assets can also be utilized from the Department of Defense to assist in moving patients from the incident site to medical treatment areas, to treat patients, and to provide diagnostic and medical support services. Volunteer personnel may also be utilized.

- **Patient evacuation.** Responsibility for movement of seriously injured or ill patients falls on ESF-8 resources. Federal assets may be utilized to help move these patients from the incident site to other facilities, or even other states. Assets that may be requested include the Department of Defense, Department of Veterans Affairs, and FEMA. FEMA can also utilize the national ambulance contract to provide transportation assets.

- **Patient care.** Assets from a variety of federal agencies and organizations are requested to assist in the triage and treatment of individuals affected by the disaster. Federal assets may also be utilized for field and inpatient hospital care, outpatient services, pharmacy services, and dental care. Patient care may also include assistance in mass vaccination and quarantine during a mass pandemic incident.

Remember that any federal assets that are called are there to provide assistance during an incident, not to take charge of the activities.

Types of federal resources. When federal resources are requested a variety of options are available, depending on what your locality is in need of. The following are federal assets that may be available for deployment, and the roles and responsibilities that the resource may undertake at an incident scene (figure 4–2). Remember that once requested, it can take from 24 to 72 hours for federal resources to arrive on scene and begin assisting in incident response.

Fig. 4–2 National Guard and other military assets may be deployed to assist in patient evacuation, patient care, and other medical functions during a disaster.

Disaster medical assistance team (DMAT). The first asset that you and your agency may need to integrate with during disaster response is the DMAT. A DMAT "is a group of professional and paraprofessional medical personnel [designed] to provide medical care during a disaster or other event."[18] With DMATs located across the country, resource availability is generally not an issue. Table 4–5 provides a list of team names and locations.

Table 4–5 DMAT names and locations

Team Name	Location	Team Name	Location
AK-1	Anchorage	HI-1	Wailuku
AL-1	Birmingham	IN-4	New Albany
AL-3	Mobile	KY-1	Fort Thomas
AR-1	Little Rock	MA-1	Boston
AZ-1	Tucson	MA-2	Worcester
CA-1	Santa Ana	MI-1	Wayne
CA-2	San Bernardino	MN-1	St. Paul
CA-4	San Diego	MO-1	St. Louis
CA-6	San Francisco Bay Area	NC-1	Winston-Salem
CA-9	Los Angeles	NJ-1	Lyons
CA-11	Sacramento	NM-1	Albuquerque
CO-2	Denver	NV-1	Las Vegas
CT-1	Hartford	NY-2	Westchester
FL-1	Pensacola	NY-4	Pomona
FL-2	Port Charlotte	OH-1	Toledo
FL-3	Tampa	OH-5	Dayton
FL-4	Jacksonville	OH-6	Youngstown
FL-5	Miami	OK-1	Tulsa
FL-6	Orlando	OR-2	Eugene
GA-3	Atlanta	PA-1	Pittsburgh
GA-4	Augusta	PHS-1	Washington DC

DMAT resources located close to the incident scene are normally not deployed, as the members are actively involved in response through their local agencies. Deployment of DMAT resources may include a full team compliment or just the personnel necessary for a specific task. The team members that deploy prepare to remain self-sufficient for 72 hours in order to avoid pulling resources from members of the community who are in need of them.

Members of the DMAT are certified as basic and advanced EMS providers, physicians, pharmacists, and more. They work from fixed or temporary sites providing a variety of medical care. While they are less likely to provide the on-scene medical care, you may find that the temporary DMAT medical facility becomes a point of transport, whether transporting patients to the facility or transporting them from a temporary facility to a more permanent medical location. As a result, you must consider that different protocols may be followed and new people are going to become part of your system.

Disaster mortuary operational response team. A second federal asset that may be requested by your agency or locality that you may interact with at a disaster scene is a disaster mortuary operational response team (DMORT). A DMORT is a group of private citizens whose expertise in a specific field assist in the ability to run temporary morgue facilities, identify disaster victims (using dental pathology and anthropology methods), and process, prepare, and dispose of remains appropriately.[19] Because many disasters have the potential of moving from mass casualty to mass fatality, as an EMS provider you may be directly involved with the movement and disposition of dead bodies to assist the DMORT. As a result, it is imperative to ensure that they are integrated into the ICS structure that is developed as the incident is mitigated.

While the above mentioned federal assets are able to provide personnel and treatment equipment, the federal government recognized a need for medical transportation assets to assist in the movement of patients from the scene to definitive medical care centers. As a result, the national ambulance contract between American Medical Response (AMR) and the federal government was written and signed in August 2007.[20] This contract allowed the federal government to request medical transport assets from AMR to areas across the country in need of assistance. According to the contract, these assets could be used for short term, though they were usually utilized for long-term deployments in an area of severe and significant need.

One of the first uses of the national ambulance contract was in 2007 as hurricane Dean approached Jamaica. On August 18, 2007, the Federal Emergency Management Agency (FEMA) activated the national ambulance contract in an effort to begin response efforts in Texas. The purpose of the activation was to ensure adequate resources were available for the potential evacuation of hospitals and care centers into areas that would not be affected by the storm. Three days after the contract activation, 300 ambulances and 25 aeromedical units were assembled at the site of the former Kelly Air Force Base, awaiting further direction on deployment. While the resources were not used as hurricane Dean did not make landfall in Texas, it did showcase the ability of AMR and FEMA to mobilize necessary resources.[21]

Conclusion

The sheer scope and magnitude of a disaster, whether natural or man-made, makes it almost impossible to respond to without the assistance of outside agencies (both private and public). With this in mind, knowledge of the available resources and their capabilities, as well as the steps in securing assistance, is important in ensuring effective response occurs in a timely manner.

References

1. U.S. Department of Homeland Security, Federal Emergency Management Agency (FEMA) (June 2007). "Robert T. Stafford Disaster Relief and Emergency Assistance Act, as amended, and Related Authorities." Washington, DC: U.S. Government Printing Office. Retrieved September 1, 2010, from http://www.fema.gov/pdf/about/stafford_act.pdf.

2. National Emergency Management Association (NEMA) (n.d.). "What is EMAC?" in Emergency Management Assistance Compact (EMAC) web site. Retrieved March 10, 2011, from http://www.emacweb.org/?9.

3. FEMA (February 2004). *Introduction to Unified Command for Multiagency and Catastrophic Incidents*. 2nd Edition, 2nd printing.

4. FEMA (August 6, 2008). *IS-775: EOC Management and Operations*, "Lesson 2: EOCs and Multiagency Coordination," in Emergency Management Institute web site. Retrieved August 20, 2010 from http://emilms.fema.gov/IS775//index.htm.

5. FEMA (August 6, 2008). *IS-775: EOC Management and Operations*, "Lesson 3: EOC Staffing and Organization," in Emergency Management Institute web site. Retrieved August 20, 2010, from http://emilms.fema.gov/IS775//EOC0103summary.htm.

6. FEMA (n.d.). "Multiagency Coordination Systems," in NIMS Resource Center web site. Retrieved August 2, 2010, from http://www.fema.gov/emergency/nims/MultiagencyCoordinationSystems.shtm.

7. FEMA (August 6, 2008). *IS-775: EOC Management and Operations*, "Lesson 8: EOC Operations," in Emergency Management Institute web site. Retrieved August 20, 2010, from http://emilms.fema.gov/IS775//EOC0108summary.htm.

8. Henrico County Community Emergency Response Team (n.d.). "Community Emergency Response Team," in HCCERT brochure. Retrieved August 12, 2010, from http://www.co.henrico.va.us/dyn/med_document/0 0000000/0000000/000000/00000/0000/800/863/hccert.pdf.

9. Community Emergency Response Team (CERT) (n.d). "About CERT." Retrieved August 12, 2010, from http://www.citizencorps.gov/cert/about.shtm.

10. Department of Health and Human Services: Office of Emergency Preparedness. "Metropolitan Medical Response System" [PDF document]. Retrieved from Centers for Disease Control and Prevention (CDC) web site http://www.bt.cdc.gov/planning/CoopAgreementAward/presentations/mmrs-oep10minbriefing-jim11.pdf.

11.   Office of the Civilian Volunteer Medical Reserve Corps. "About the Medical Reserve Corps." Retrieved from http://www.medicalreservecorps. gov/About.

12.   FEMA. "The Disaster Process and Disaster Aid Programs." Retrieved from http://www.fema.gov/hazard/dproc.shtm.

13.   FEMA. "Glossary/Acronyms," in NRF Resource Center web site. Retrieved August 22, 2010, from http://www.fema.gov/emergency/nrf/ glossary.htm#N.

14.   FEMA. "Emergency Support Function Annexes: Introduction" [PDF document]. Retrieved from http://www.fema.gov/pdf/emergency/nrf/ nrf-annexes-all.pdf.

15.   FEMA. "Emergency Support Function #8—Public Health and Medical Services Annex." Retrieved from http://www.fema.gov/pdf/emergency/ nrf/nrf-esf-08.pdf

16.   FEMA. "Glossary/Acronyms" in NRF Resource Center web site. Retrieved August 22, 2010, from http://www.fema.gov/emergency/nrf/ glossary.htm#E.

17.   U.S. Department of Health and Human Services (11 March 2009). "Office of the Assistant Secretary for Preparedness and Response (ASPR)," in Public Health Emergency web site. Retrieved June 12, 2011, from http:// www.phe.gov/about/aspr/Pages/default.aspx.

18.   _____. "Disaster Medical Assistance Team (DMAT)." Retrieved August 13, 2010 from http://www.phe.gov/Preparedness/responders/ ndms/teams/Pages/dmat.aspx

19.   _____. "Disaster Mortuary Response Teams (DMORTs)." Retrieved August 15, 2010 from http://www.phe.gov/Preparedness/responders/ ndms/teams/Pages/dmort.aspx

20.   U.S. Department of Health and Human Services. "FEMA National Ambulance Contract" [PDF document]. http://www.njha.com/ep/ pdf/772010101922AM.pdf.

21.   FEMA. "Ambulances staged in Texas for Hurricane Dean" [photo]. Retrieved from http://www.fema.gov/photolibrary/photo_details. do?id=31663.

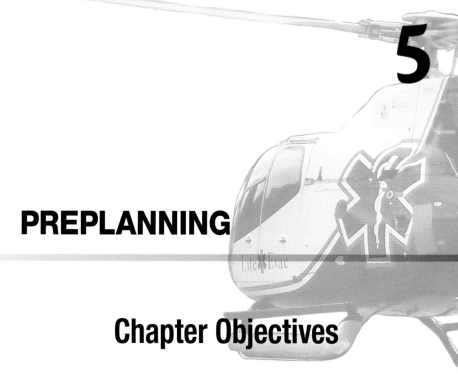

PREPLANNING

Chapter Objectives

Upon completion of this chapter, readers will be able to:

1. Understand the need for preplanning

2. Understand the basic preplanning process

3. Recognize areas in their locality that may benefit from EMS preplanning

4. List suggested elements in an EMS preplan process

5. Apply everyday preplanning to large-scale events

Scenario: After picking up an overtime shift, you have been assigned to a medic in a response area that you are unfamiliar with. After introducing yourself to your partner for the day and checking out the truck, you sit down to some breakfast when you are alerted to a call at a local manufacturing warehouse for a patient who is unconscious after seizing.

"Medic 5, you are dispatched for a 33-year-old male who is unconscious and seizing. The caller is advising that you should park in loading bay B for quickest access to the patient."

You turn to your partner for some guidance, since you are unfamiliar with the facility and the area.

"I can get you to the building, but I'm not sure where loading bay B is. We can stop at the front security gate and ask for directions once we get to the facility," your partner says.

You cannot help but wonder what else may happen with this shift if this is how it is starting.

Knowing you are stationed in an area that you were unfamiliar with, what is your responsibility for becoming familiar with the response area? What could you and your partner have done differently to improve the response time and efficiency?

Introduction

Knowing that you cannot memorize every building within your response area and mutual aid areas, how can you ensure that you have access to the necessary information for adequate response? The simple answer is to preplan. Firefighters are no strangers to preplanning. In some departments each shift is required to complete preplans during their on-shift hours, while other departments may provide a list and hope that members complete it. However, preplanning is normally completed with a focus on the fire aspects of the building (electrical shutoffs, water connections, sprinkler systems, etc.) and little consideration for the information that may impact EMS response to the facility. EMS providers should begin to take the initiative to conduct and document preplans.

Preplanning means to plan ahead. Preplanning allows you to identify target hazards, identify limitations (of your crew, your agency, and your local resources), and access accurate information when needed. This process provides information that can aid in decision making during response operations. The primary purpose of a preplan, whether fire or EMS based, is to improve operational efficiency and effectiveness. Incident command and scene management personnel will use a preplan to determine staging locations, incident approach, incident setup areas (e.g., treatment, transport, etc.), and determine the best approach for incident mitigation. Another use of the preplan is for training. It is beneficial to write a plan, but even more beneficial to train with the plan. Training on plans helps you become familiar with the information, which makes it easier to recall during the incident. Training with preplans can be as simple as roundtable discussions about potential issues or as complex as on-site building familiarization drills including appropriate apparatus placement.

You may also find that the preplan is a great way to share information. Initially, preplans should be shared with other shifts and other stations that have an expectation of response to the facility. This ensures that everyone is using the same approach and has the same information during response. Preplans can also be used to work with other localities to ensure that any mutual aid response is aware of your plans. It provides an open line of communication to all involved. Finally, and sometimes the most overlooked, is the use of the preplan at the emergency scene. Often times, responders may forget that the preplan has been written and fail to refer to it during development of the response and operational tactics. Writing the plan and training with the plan is one thing. Using the plan at the appropriate time is another.

Developing a Preplan Template

In a department that does not conduct preplanning, beginning with a standard template can make the preplanning process easier. A format ensures that every preplan contains the same information and information is easy to find. Every preplan should contain the same basic information and components, though changes will be necessary based on the specific needs, design, and functionality of the facility. Preplans can be basic or they can be complex. Regardless of the format though, the first page of your preplan should contain the information you will need during the initial stages of your response, from dispatch to the minutes after you arrive on the scene. The information contained beyond the front page assists in the development of operational strategies beyond the first few minutes of incident response. While format is not necessarily important, the minimum information obtained should be consistent.

Example EMS preplan format

EMS PREPLAN FORMAT

Basic Information

Occupancy Name: _____

Occupancy Address: _____
(Street)

_____
(City)   (State)   (Zip Code)

Point of Contact: _____
(Name)

Phone Number: ( ) _____ Fax Number: ( ) _____

E-mail Address: _____ Cell Phone: ( ) _____

Occupancy Type: _____

General Directions to Occupancy: _____

_____

_____

Response Companies   1ˢᵗ: _____ 2ⁿᵈ: _____ 3ʳᵈ: _____

Building Information

Hours of Operations: _____ **Number of Occupants:** _____

Number of Exits: _____ **Number of Floors:** _____

Number of Elevators: _____ **Number of Stairwells:** _____

Onsite Medical Personnel?   Y _____   N _____

Onsite Medical Equipment?   Y _____   N _____

Special Hazards: _____

Additional Notes

_____

_____

_____

Maps

Plot Plan
(Use this space to draw the plot plan for the facility. If a plot plan is available, attach it in place of this page.)

MAPS

Building Plan
(Use this space to draw the building plan for the facility. If a building plan is available, attach it in place of this page.)

Maps

Floor Layout Plan
(Use this space to draw the floor layout plan for the facility. If a floor layout plan is available, attach it in place of this page.)

ADDITIONAL NOTES/INFORMATION

Use this page to document any additional information that you have not already gathered that is pertinent to response to this facility.

_____

_____

_____

_____

_____

_____

_____

_____

_____

_____

_____

_____

_____

_____

_____

_____

Basic information

The most basic information that should be obtained on every facility is the occupancy information. Occupancy information includes the name of the business, physical address of the business, any pertinent telephone numbers for the business, and a primary point of contact (POC). The primary POC is normally someone who can provide the most information. This may be a business owner, facilities manager, or the landlord. Along with the basic business information, basic response information should be included. Response information may include 1st, 2nd, and 3rd due agencies (fire and EMS). This ensures that anyone referring to the plan during an emergency incident is familiar with which companies and what resources are available. Every preplan should also include a basic set of directions to the facility from the station. This may include any major intersections or special considerations for response, such as low wires, low clearance, and road construction. The directions will obviously need to be changed based on which station is viewing the preplan, and while they do not need to be specific, they should provide a general overview.

Maps

Maps are another important piece of information that should be included in every preplan. Maps provide not only an idea of building layout and facility layout, but also detailed information about the layout of each floor and information on where large numbers of people may gather within the facility. The first map that should be included is a building layout map. The building layout map provides information on the facility itself, including entrances and exits, use of space (e.g., office space vs. warehouse space), potential incident hazards, and systems that may impact response (i.e., electrical system, fire suppression system, medical supply area). The building layout map should also include areas of large congregation and areas of safety concern, as these may cause large patient load.

Another map that might be considered is a floor plan map. If the floors of a facility are similar, you may only need one floor plan map. However, if floor plan differs significantly between floors, additional maps may be needed. A floor plan map provides information on areas of rescue, entrances and exits (which may provide a basis for where patients will be located), stairs and ramps for patient extrication, and any potential medical equipment/treatment areas that may be of benefit during incident response to the facility. For instance, many high-rise buildings have stair chairs at the landing of a staircase. Knowing this prior to incident response may save you time in trying to bring equipment in to assist in patient evacuation.

In addition to those maps, consideration should be given to including a plot plan. A plot plan is a basic drawing that provides an outline of the facility and the surrounding land. Figure 5–1 provides an example of a basic plot plan. A plot plan shows surrounding buildings, roads, parking lots, and potential obstacles for incoming apparatus. The plot plan can also show areas that are inaccessible to facilities due to the obstacles or lack of road access, such as the rear of a facility. The plot plan can be used to designate predetermined staging areas or access and egress routes for transport trucks during a mass casualty. This information will assist in making tactical decisions during the initial phases of an incident.

FEMA recognizes the following facilities as target hazards:

- Nursing homes
- College dormitories
- Penal institutions
- Public assemblies
- Enclosed malls
- Bulk storage facilities and tank farms
- Mill buildings
- High-rise buildings

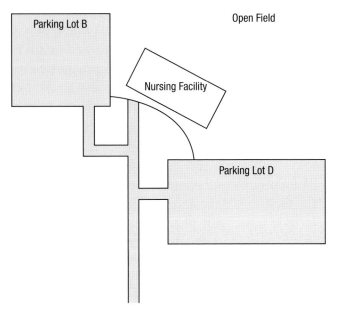

Fig. 5–1 Plot plan

The Next Step

Once a format for a preplan is developed, the next step is determining which facilities in your response area should be preplanned. In an ideal world, you could preplan any and every facility with the potential to stress your response system. However, with the limited resources, time, and funding facing EMS providers, there is a need to prioritize which facilities are preplanned. The Federal Emergency Management Agency (FEMA) categorizes facilities within a locality based on use, size, and occupancy.[1] The categories used in this program are also beneficial when prioritizing facilities for EMS preplanning activities. While there are a large number of target hazards, in this chapter the focus is on nursing homes, college dormitories (school facilities), public assemblies, enclosed malls, transportation buildings, and high-rise facilities. These locations have a high probability for high patient load and should be considered priorities in the EMS preplanning process. Once you have determined which facilities in your response due will be preplanned, you can begin the process. You should collect the basic information, as previously discussed in this chapter, but when gathering information for the preplan, you should consider the special issues that arise with each type of target hazard.

Nursing homes/assisted living facilities

A nursing home is considered a target hazard by virtue of the age and special needs of the population that resides at the facility. The higher age of the residents means that you may be forced to deal with a large population who are not only impacted by the emergency incident itself, but also have secondary medical conditions that must be treated during the incident. Determine whether the facility houses those who are fully dependent on medical staff or is a mixed population (a facility with independent and medically dependent residents). Your focus during an incident may need to be on those who will need assistance in evacuation, so ask where each type of resident lives within the facility grounds. You should also recognize the need for specialized medical treatment once residents have evacuated the facility. In a day-to-day situation, residents have access to ventilators, oxygen, suction, IVs, and other medical equipment that support them. However, once evacuated, access to this equipment becomes limited. With that in mind, preplanning operations may include immediate request for support medical equipment as necessary. Additional information that may assist in the preplan of this target hazard is the point of contact (POC) for medical information regarding the patients. This individual may be different from the POC for the facility itself. Most nursing homes have medical support staff. These individuals will understand the specific medical issues of the residents and will serve as a great source of information during incident operations.

College dorms/school facilities

College dormitories and, in general, school facilities, raise several important considerations when preplanning incident operations. As has been increasingly seen in incidents in recent years, schools are a significant target hazard. Because of the population and media attention that is brought on by events on campus, these facilities have also become a potential target for terrorist acts. When preplanning school facilities, keep in mind that there are a large number of special considerations that should be included within the preplan document.

After collecting the basic information, the next piece of information that you should determine is when the population of the facility is at its highest. It is important to note in the preplan what the school year and school hours are, as the population will significantly decrease when school is not in session. A high population increases the probability of a high patient load. You should also determine which areas of the campus house classes, administrative offices, dormitories, etc. Regardless of the type of school, there will be areas on-site that hold a greater number of students/visitors/staff. These areas may require a larger number of resources should the incident be located there. Other areas may require less manpower and resources to mitigate a situation. Knowledge of the location of these areas can help with tactical and resource allocation decisions.

Another important part of the preplan of a school facility is determining the educational focus of the facility. Schools that provide general studies may have less potential hazards, while technical or trade schools may house chemicals and supplies that can pose a higher medical and fire safety threat. This is important to know so that you can immediately begin considering the need for hazmat, fire, and EMS resources. You should also consider whether the facility is a training facility for a special needs population (e.g., blind, deaf, mobility impaired). Regardless of the focus of the facility, you should always determine the POC who is able to provide you the most accurate information on the student population. Patient tracking and accountability may be easier to maintain at a primary school. At primary schools (those from kindergarten through grade 12), this POC may be the school nurse, as he or she is more familiar with the students' varying medical issues. When conducting the preplan, also remember to work closely with the school administrator to determine what policies are in place to determine any missing students during emergency incidents.

Public assemblies

By definition, a public assembly is a facility in which a portion or all of the facility is used for gatherings of 50 or greater people "for such purposes as deliberation, worship, entertainment, eating, drinking, amusement, or awaiting transportation." Given this definition, it is safe to assume that public assemblies cover most of the businesses and buildings within your response area. In fact, most of the target hazards discussed in this chapter should also be considered public assemblies.

Because of the broad range of facilities that fall under the title public assembly, there are a large number of general and specific considerations that should be kept in mind as you preplan the locations. The population found at public assemblies can be so wide ranging that planning for the medical issues to expect at an incident may be much harder than for other locations. Therefore, it may be easier to preplan based on the general layout and function of the facility you are responding to. In planning for response to public assemblies, keep in mind the general characteristics that may impact operational activities and tactical planning.

- Large unbroken areas support the potential for large numbers of people being injured in one area of the facility.

- Open stairwells may assist in the movement of personnel and evacuation of patients out of the facility during an emergency incident. Knowledge of stairwell locations is imperative for quick movement throughout the facility.

- Doors that are often locked for security reasons may initially delay access to some areas of the facility. Having knowledge of which doors are secured may assist in preplanning actions to ensure access through these doors.

- Occupancy flow that is designed for entering and exiting from the same point can hinder access to the scene and evacuation of those in the facility, especially if responders are trying to enter the same area that victims are trying to exit. The funneling effect created by guiding people to use the same point as the entrance and exit can create a logjam and slow access to those individuals who need help. This can also influence your initial actions. If people rush to evacuate during an emergency incident they may create a stampede environment and cause injuries, which normally occur right around the exits. If this happens, your first action will be to care for and evacuate those people first.

- Single access/egress roads to a facility can cause traffic flow problems for emergency vehicles attempting to access the scene of an emergency incident. Consistent access to the scene is imperative. While preplanning the facility, review access and egress points. Initial tactical decisions may include immediate involvement of law enforcement for traffic control. Placement of staging may also be dictated by the ease of access to the facility.

Public assembly facilities include, but are not limited to:

- Assembly halls
- Auditoriums
- Bowling alleys
- Churches
- College/university classrooms
- Conference rooms
- Courtrooms
- Dance halls
- Drinking establishments
- Exhibition halls
- Gymnasiums
- Libraries
- Movie theaters
- Museums
- Passenger stations and air terminals
- Performance theaters
- Restaurants
- Skating rinks

Medical facilities

While not specifically listed as one of the target hazards by FEMA, medical facilities present a large number of special concerns when responding to an event at the facility. Medical facilities can include dialysis centers, outpatient surgery centers, and hospitals. Each of these facilities present a special need based on the population, the equipment, and procedures being carried out on a daily basis. Preplanning of medical facilities should include collection of the basic information, but also an in-depth review of information specific to the facility itself.

Because of the nature of the facility and the licensing requirements to run medical facilities, most have detailed emergency response plans. When conducting your preplan, ask if you can have a copy of the plan to include in your preplan. The facility's emergency plan can provide you with information that can assist in making tactical decisions. One of the tactical decisions that may be affected by a facility's emergency plan is whether or not to shelter in place. Many facilities have fire doors or facility protection that allows for patients and visitors to be kept safe if they are sheltered in place. Not only does sheltering in place minimize the number of people who need to be evacuated, but it also ensures that the patients have access to medical equipment that might not be immediately available once evacuated.

Also, when gathering maps of the facility, make sure that you determine the specific layout. Determine where the most vulnerable populations are. These individuals most likely require the most medical intervention in a treatment area, and are the most likely to suffer from medical problems as a result of an incident. Remember that most medical facilities are host to a broad variety of individuals who have been diagnosed with a variety of medical problems. Along with the patients and staff at medical facilities though, you must also consider the visitors to the facility. While preplanning the medical facilities you should determine how they track visitors. This information can help in determining when a facility has been fully evacuated and how many people you should account for during the incident.

Each of these locations has a large potential for patient load and should become priority in conducting and completing preplan activities. These locations also have special considerations that should be kept in mind while conducting preplan activities to ensure that the most adequate, accurate, and useful information is collected for use during your tactical decision-making.

Conducting the Preplan

Once you have determined the priority facilities for preplanning, you must determine the steps you want to take to gather the information. The first step is to determine your preplanning partners. As previously stated, information from a preplan will be shared with other agencies, and you should also turn to other agencies within your locality to determine whether the information you need has already been gathered. Partners in preplanning can include but should not be limited to the following agencies, organizations, and people:

- Fire department personnel (fire marshal's office)

- Public works department

- Police department

- Hospitals

- School board

- Facility and building management

Not only do these partners potentially have information that you require, but they may have already conducted preplans on the facilities you are looking to preplan.

After speaking with these partners, you should have a better understanding of what information you still need from the facility. Once you have made that determination, you should contact the facility directly to schedule an interview with the point of contact. Conducting an interview will allow you not only to ask the important questions, but also give you a face-to-face meeting with the person you will be interacting with at an emergency incident. When you meet with your point of contact, you should explain the benefits of the preplan. After gathering the necessary information and requested supporting documents (e.g., maps, emergency plans, etc.), you should ask to conduct a walkthrough of the facility. This will allow you become familiar with the facility and make note of any important findings. You may also be able to use the walkthrough as an opportunity to develop the plot plan, noting the building location on the grounds as well as access and egress routes.

Writing the Preplan and More

Information is important to have, but if you do not write it down and share it, then it only benefits you. After gathering the information for your preplan, compile the information into the predetermined format. Any additional considerations and concerns that were raised during your preplan interview should be addressed and mentioned, and appropriate maps should be included. It is also important to ensure that the information is accessible when needed and that it is provided to the appropriate individuals. A preplan locked inside an officer's desk will be no help during response and emergency operations. A preplan that is placed in a location on the apparatus that is accessible to everyone provides the best support to responders. The preplan must also be shared with the mutual aid companies. These companies will make some slight changes to the documents (directions, response companies, etc.), but the information on layout, plot plan, etc., will remain constant and ensure that incoming agencies have the same information.

Once written and shared with emergency responders, you must also train on the preplan. As with training on ICS, you should also train on emergency response utilizing the preplan. Training can be as simple as a review of the preplan and apparatus placement expectations or as complex as practicing unit placement at the occupancy. Training also allows for review of the preplan to ensure that unit placement and other predetermined operational decisions are the best action to take during response to that specific occupancy. Training can assist in making changes to the preplan to ensure it is as effective as possible.

Large-scale events that should be considered for preplanning include:

- Fairs

- Craft shows

- Concerts

- Celebrations

- Sporting events

- Anniversary of historical events

- Inaugurations

Preplanning Considerations for Large-Scale Events

It seems unassuming at first. A craft fair, a car show, an outdoor concert, or one of many other types of activities is scheduled within a locality (figure 5–2). Notification to the fire department, EMS agency, and police department may occur initially, but is not maintained during the planning process of the event. Large-scale events can, in some cases, create a "mini city" within your response area. Preplanning for the everyday incident is an important task and should be the primary focus of your preparedness activities for emergency response planning. However, also consider the need for preplanning large-scale events in your locality. Large-scale events can be classified into two categories:

1. Regularly scheduled events. These are events that usually fall on the same date, month, etc. There are often records and historical data that can provide information to assist for planning, and the plans can normally be carried from one year to another with only minor changes. This means that the events are normally easy to plan and prepare for.

2. Special events. These events are not regularly scheduled and may only be a onetime concern. They fluctuate in time, place, purpose, etc., and present a special concern when planning. There is no historical information on the event, so planning for potential problems and issues must be based on research and speculation.

Regardless of the type of event, preplanning is a necessity to ensure that your agency, your locality, and the participants of the event are best prepared should something occur.

Fig. 5–2 An outdoor craft fair in the middle of summer has a significant potential of becoming a large-scale event.

Beginning the planning process

Events you have no notice of are hard to preplan and will normally be dealt with as incidents arise. They normally require mutual aid resources to ensure that daily 911 calls as well as issues that arise at the large-scale event are adequately handled. Events that you have knowledge of are easier to prepare for. While you may not be the lead individual for the planning process, you can ensure that the appropriate steps are taken to prepare for the event. The minute you receive notification of an event you should begin the planning process.

The first step in planning for a large-scale event is developing a planning team.[2] The planning team should be multi-disciplinary, representing any public or private organizations that are impacted by the event. At a minimum, the planning committee should include

179

representatives from emergency management, law enforcement, fire, rescue, public works, public health, and transportation. These agencies represent a large portion of the resources that are needed to support large-scale events. The event coordinator should also play an integral part of the planning processes, as this person may be able to provide information on expected participants, desired outcome, etc. Other agencies may be necessary, depending on the type of event, including animal control, local legal aid, community service organizations (Red Cross, Shriners, Lion's Club, etc.), and possibly representatives of the school board.

The initial planning meeting should focus on the mission of the committee and should serve as a chance to gather general information about the event and the event's participants. One consideration that must be kept in mind during planning is the purpose of the event and the type of activities. If the event includes thrill rides and physical activities, you may be faced with an increased number of patients suffering from traumatic injury. If the focus of the event is the enjoyment of performers, musicians, etc., then your focus may need to be on the potential for medical-based issues. The impact on the locality also needs to be considered during the planning process. Planning should ensure that local emergency medical services needs are able to be handled during the event. Incidents at the large-scale event should not affect the local citizens' ability to receive EMS care. For this reason, mutual aid resources will become vital in ensuring that the needs of the event and of the locality are met. The initial planning meeting also provides an opportunity to begin a hazard analysis of the event. This will ensure that all potential hazards are considered and plans are written accordingly.

Hazard analysis

A hazard analysis is a systematic decision process designed to recognize what may occur and the impact of the events on a community or event. Because of the nature of large-scale events, a hazard analysis of your event should be broad in scope, providing for the potential of a variety of events that, while historically may not have presented as an issue, have the potential to occur during the event. Conducting a hazard analysis is a four-step process: identify, profile, prioritize, and plan.

Hazard identification. This step involves review of the potential hazards associated with not only the event, but the local area. Remember to consider the possibility of all types of hazards, including man-made (criminal/terrorist events) and natural. In order to determine the area's potential hazards you should review current hazard plans, historical data regarding similar time periods and similar events, and, most especially, hazards that may be unique to the event. Also remember to consider secondary hazards that are a result of the event. For example, if you are in a planning session for a big air show, you should identify the potential of an airplane crash during the event. If food is being prepared and cooked at the event, burns and fire hazards should be identified as a potential issue, as could a significant outbreak of food poisoning due to improper food storage/preparation. Remember, just because a hazard may not have been a problem before does not mean it will not be a problem for this event.

For most events there are two significant potential hazards. The first hazard that should be considered for all large-scale events is weather related. Since weather can frequently change, there is a necessity to monitor weather trends for the time of year as well as the forecasted weather for the event time frame. Weather will have a major impact on the number of participants, as well as an impact on the number of incidents that occur during an event. Consider the probability of seasonal weather issues as well as the time of day of the event. For example, thunderstorms normally occur in the afternoon hours, so a morning event has a lower hazard probability. Some weather events may also require a monitoring and alerting system be put in to place for the event, such as those issued by the National Weather Service.

The second major hazard that should be considered is a potential for a criminal or terrorist act to occur during the event. A large-scale event presents a tempting opportunity for a terrorist or criminal act because it allows for an impact on a large number of people. Political, religious, environmental, and racial events are also targets for terrorist or criminal threats. The primary responsibility for planning for response to a terrorist or criminal act falls with local law enforcement. Research and intelligence provided by law enforcement agencies can assist you in determining not only the chance of the event occurring, but also determining the potential threats associated with the event (e.g., toxic chemical or bombing incident). Planning for the potential terrorism hazard allows you to work with other responders to ensure responder readiness (including training and equipment). Determining response duties in a terrorism event prior to the incident can also assist in ensuring a smooth and efficient response.

Profile identified hazards. Writing a hazard profile may be the most time consuming part of the hazard analysis. Profiling hazards begins with determining the magnitude of the hazard. This includes not only the size of the hazard, but also the area that the hazard can potentially affect. At a large-scale event, weather hazards can affect the entire event, while a fire hazard may be limited to specific areas of the event. You should also profile the frequency of the hazard, including any potential seasonal patterns associated with the hazard potential. For example, certain weather events increase in frequency potential during specific times of the year (e.g., hurricanes from June through October). This means that when "off-season," the potential for that hazard is much lower. You should also consider the duration of each hazard. Events that are of short duration may require only minimal support necessary for response. A final considering in profiling a hazard is speed of onset. This is most important in determining the amount of time you have to provide a warning to event participants. A greater warning time is significant because it may decrease the amount of injury and loss of life during a hazard.

Prioritizing risks. FEMA defines risk as "the predicted impact that a hazard [has] on the people, services, and specific facilities at the event and in the community." The ability to determine the risk of each hazard allows you to focus your planning efforts and resources on those hazards posing the highest risk. Determining vulnerability for each hazard can be done by reviewing your locality's emergency operations plan (EOP). If the EOP focuses a significant amount of attention on planning for that hazard, it should be considered a great risk. Whether or not critical facilities such as hospitals and emergency service facilities are impacted by the hazard should also be considered in determining the risk. Risk prioritization is also impacted by the delays associated with receiving assistance from response. These can range from delays caused by simple traffic and road debris to such complex issues as arrival of responders who are trained to handle the specific hazard. Once risk has been assessed, response priorities should be determined.

Regardless of the hazard, the first priority should be life safety. Life safety focuses on high-risk populations, people in the hazard areas, and potential search and rescue missions. The second priority is return of essential functions, including a focus on the fact that emergency response assistance is unavailable if the emergency facilities (fire, police, and EMS stations) are impacted by the hazard. The third priority is critical infrastructure. Once life safety has been ensured, the restoration of critical infrastructure resources such as communications, essential utilities, and transportation is key to maintaining life safety and function of the community.

Plan for vulnerabilities. Once you have identified and prioritized potential hazards, you must plan for hazards with the highest risks. Planning can be scenario based, creating a potential timeline for hazard development from initial warning to actions that need to be taken to resolve the hazard. Writing a realistic scenario involving hazard impacts will allow you and the planning committee to determine when local, regional, and even statewide mutual aid may be needed to support hazard response.

Additional planning considerations

In addition to planning for issues associated with weather, criminal activities, or other hazards determined to be a potential, you should also focus on the high potential of "minor" incidents occurring during the event. Minor incidents may include bee stings, difficulty breathing, sunburns, headaches, and cuts and scrapes. While you may tend to focus planning for the major events, do not forget that it is the minor issues that will be most prevalent throughout the large-scale event.

Staffing. Staffing is a major concern when planning for the large-scale event. In order to ensure local citizens are not greatly impacted due to decreased capability of response to calls, you must provide additional staffing to the event. The question exists though in how to determine the number of providers needed, given the size of the incident. In its special events planning program, FEMA provides assistance in determining the staffing levels to be considered during a large-scale event. Table 5–1 can be used to determine staffing levels based on the number of expected event participants. For example, in an event with 5,000 people, approximately 20 first aid personnel should function at the event. The table also has information about the expected injuries given the size of the crowd. At that same event with 5,000 people, you should plan for at least 15–65 attendees experiencing a medical emergency. These numbers do not reflect the number of patients to expect should an emergency incident occur at the large-scale event. Planning for mutual aid staffing of emergency incidents, such as bad weather, structural collapse, fire, etc., should be significantly higher.

Table 5–1 Staffing and planning considerations for large-scale events

	500	5,000	50,000	100,000	500,000
Security Officers	1	10	100	200	1,000
Medical Aid Post(s)	1	10	100	200	1,000
Medical Aid Personnel	2	20	200	400	2,000
Persons with Medical Injuries	1.5–6.5	15–65	150–650	300–1,300	1,500–6,500
Parked Vehicles (1 per 3 spectators)	167	1,670	16,700	33,400	167,000

Medical facilities. Once you have determined staffing needs and levels, you must also plan for the location of the personnel and the best way to observe and assist event participants over an area that can be small, such as a craft fair, or one that may cover several acres, such as a state fair or concert. Having the people is not enough. You must decide the best way to utilize them to provide necessary care throughout the event. One suggestion is the use of strategically placed medical aid stations (figure 5–3). Medical aid stations are able to provide a minimal level of medical intervention (bandages, ice, etc.). They provide a respite from the weather and allow for basic medical assessment and treatment to be conducted. FEMA also developed recommendations regarding appropriate numbers of aid stations during large-scale events. These numbers, included in table 5–1, provide a suggested number of aid stations to best assist event participants. In an event with 50,000 expected participants, FEMA recommends approximately 100 aid stations. This table can assist you in understanding logistical needs for each of these locations, as well as the steps necessary to assist in obtaining the appropriate staff levels as recommended by FEMA.

Fig. 5–3 Medical aid stations are able to provide minimal level of medical intervention at large-scale events.

Along with fixed aid facilities, you should consider the use of mobile aid to provide coverage to those areas that have no medical aid stations but do have participants who may be in need of medical aid. Setting up mobile assistance can mean the use of multiple transportation methods to assist in accessing those in need of aid. Mobile medical aid can include walking teams, bicycle teams, cart teams, and even motorized personal transporter teams. When determining the best way to set up mobile medical care, you should remember to look at the layout and terrain of the event space. For events in stadiums with narrow walkways, walking teams may be best method to access patients. Events that cover significant space and have rougher terrain (dirt roads, rocky pathways, etc.) are best served by carts or gators that allow for off-road access to patients.

Mutual aid resource considerations

During a smaller incident, your locality may be able to handle the calls for medical assistance with local resources. When those resources become overwhelmed, mutual aid becomes a necessity for handling both the incident and the everyday 911 calls that will continue to be received. With a planned large-scale event, mutual aid becomes a necessity to ensure that the appropriate medical staffing levels are maintained. The use of mutual aid resources creates more potential planning issues. The first issue for consideration is medical protocol. In EMS, protocols are essential. When resources are brought in from outside the local response area, protocols of each agency may conflict with what you are used to. A review of state regulations must be conducted and a determination made about the appropriate approach to the protocol issue. Will all providers act under the same protocols (those of the host agency) or will each agency utilize their own protocols during response? These are questions that must be answered. You must also consider the issue of equipment. Not only will agencies have different equipment, but they may carry different drugs for patient treatment as well. Plans must be made to prepare for the issue of equipment and drug exchange after transport of patients to medical facilities, whether on- or off-site.

A second consideration when utilizing mutual aid resources is the need to provide directions to area facilities. Whether in a disaster or a preplanned large-scale event, ensuring a smooth transition from patient transport to the off-site medical facility is key. Mutual aid resources may not have knowledge of the local facilities, and will require guidance on movement from the incident to the medical facility. In preplanning for the large-scale event, predetermine the medical facilities that will be utilized by transporting units. This will allow you to gather the appropriate information and develop a guide or handbook for mutual aid resources assisting at the event. Information that should be provided on each hospital includes facility name, physical address, phone number and radio frequency (for hospital reports), directions from the event, a map to the hospital, and a diagram of the hospital. This information will allow mutual aid resources to have knowledge of the facilities and be prepared for smooth transportation from the incident to a definitive care facility.

Scenario. You and your partner are finishing lunch when dispatch alerts you to a call.

"Medic 3 and Engine 2, respond to 1001 Smith Drive to the Christ Church for a difficulty breathing."

The dispatcher states that the caller is in the main sanctuary. You recognize the address as a large church that has multiple buildings on the campus. While you know the physical location of the church, you are uncomfortable with the actual location of the patient. You remember that you conducted a preplan of the facility a few months back and grab the notebook in the center console. After looking at the plot plan, you realize the main sanctuary is behind the first building when you enter the grounds and that the main entrance is on side D of the building. You let your partner know that it is quicker to enter from the Smith Street entrance as it passes directly in front of the main sanctuary.

As you pull up to the main entrance of the building you are met by church members who are screaming that you need to come quick. You quickly glance over your shoulder looking for the engine when you hear the engine contact you over the radio.

"Engine 2 to Medic 3, we are having trouble locating the building, can you provide us with some directions?"

"Medic 3 to Engine 2, if you are coming from the Smith Street entrance, we are the second building on the left, you will see the ambulance in the parking lot."

"Engine 2 copy."

As you enter the building, you approach your patient and determine she is unconscious, apneic, and pulseless.

"Medic 3 to Dispatch. Be advised we are working a full arrest."

As you release the microphone, Engine 3's crew enters the room and you begin to treat your patient and prepare for immediate transport. After transporting the patient to the nearest hospital you sit down to write the report and cannot help but reflect on the call.

- What would have happened if you had not been familiar with the call location?

- How long would your response have been delayed if you had not referred to the preplan to determine the building location and quickest access to the building?

- Did Engine 3 have a copy of the preplan? If yes, did they choose not to refer to it while en route to the incident?

You realize that while you cannot change the outcome of this call, you can prepare for future calls both at this location and at others that may benefit from a thorough preplan.

Conclusion

Preplanning can mean the difference between immediate access to a patient and having to wait for a staff member to direct you to the patient. It can provide for a smooth transition from arriving on scene to transitioning to large-scale response activity. A properly conducted preplan can provide immediate access to information that will assist you through the whole call, especially during the initial phases of response. Even more importantly, a preplan that is shared with partner agencies assists in creating a coordinated response to emergency incidents. Working and training with other agencies using the preplan ensures that each response agency is working cooperatively from the same plan and hopefully leads to a quicker resolution. Take the time to develop a strong preplan and, most importantly, make sure that your agency utilizes the information gathered.

References

1. U.S. Department of Homeland Security, Federal Emergency Management Agency (FEMA) (December 2009). *Command and Control of Fire Department Operations at Target Hazards*, 1st Edition, 7th Printing.

2. FEMA (July 2010). *Special Events Contingency Planning*, "Lesson 2: Pre-Event Planning," in Emergency Management Institute web site. Retrieved July 30, 2010, from http://emilms.fema.gov/is15b/SEP0102summary.htm.

TRAINING AND EXERCISE

Chapter Objectives

Upon completion of this chapter, readers will:

1. Understand the benefits of practice exercises

2. Recognize the process for conducting a needs assessment

3. List the types of exercises

4. Understand the process of developing and conducting an exercise

5. Recognize the process of developing and conducting an after action review

6. Understand how to utilize information from the after action review to improve policies and procedures

Scenario: Your captain announces that today everyone in the station will attend ICS refresher training. It should only last for an hour and a half.

An hour and a half is not that bad, you think to yourself after his announcement. At least after the training you will be able to come back and catch the college football game, as long as there are no calls.

The captain goes on to say that after the classroom portion of the training everyone will go out to the training ground and run through some practical exercises, including some burn building and vehicle extrication drills.

"So much for our day of rest," you mutter under your breath to your partner.

"Why do we have to do that?" you think. "Isn't an hour and a half in the classroom boring enough? How exactly will working through scenarios on the whiteboard help us at an actual incident?"

What is the benefit of follow-up training once you have received your initial certification? How can you tell if things are working or if other training methods are best for you and your fellow responders? Whose responsibility is it to ensure everyone is trained appropriately?

Introduction

You know the drill. Once a topic is taught, the expectation is that you will remember the information and use it for the rest of your service. Many topics in the emergency services field require constant review and updates as well as application of the information to ensure that you maintain the most up-to-date information about the incident command system structure. Training in any field of emergency response is not a one-time thing. It needs to continue, not only through field use and experience, but through every method possible to ensure that you are fluent in its use. Procedures and practices that you only utilize when absolutely necessary will never be effective and efficient. With that in mind, review the potential training methods available to you and your fellow responders.

Initial Training

Basic training is important to ensure that you are prepared to serve in the emergency services field. Basic initial certification for most agencies includes at a minimum CPR, EMT, and now, because of federal requirements, basic NIMS training. The normal medium for providing basic training is through classroom lecture. Application of ICS principles commonly occurs through in class discussions or "whiteboard" scenarios. During these training sessions you work through the scenarios on a board, talking or writing out most of your actions instead of carrying them through. While effective, there is minimal hands-on practical training to assist in reinforcement and application of principles taught during lectures. An increasingly popular and effective alternative training medium is computer based training (CBT) programs. CBT programs provide many benefits. First, they are cheaper to conduct because they do not require travel, lodging, or instructor expenses. As with classroom training, CBT programs allow for participation in scenarios. One problem with CBT programs, though, is that they are taken by one person at a time. They do not allow for face-to-face interaction with other students to exchange ideas, thoughts, and information.

Standard practice in initial certification programs is that students are presented the information and complete the course, some with certification testing, others with certificates of completion. The students then go to the field to apply the knowledge to emergency calls. Departments may conduct continuing education, but participation may or may not be mandatory, and topics may not always be relevant to the biggest issues facing members of your department. Initial training is designed to provide the basic "book" information, with the hope and expectation that students will pursue additional training and certification to strengthen their skills. As a provider, you are responsible for completing the initial training as set forth by your organization, but you should also set expectations for yourself for completion of additional training. Continuous training on the application of the ICS structure is necessary to ensure appropriate and efficient application on an emergency scene. However, in order to get personnel thinking critically about the information shared in training programs, exercises should be conducted.

Practice Exercises

Training and application of ICS principles in a training environment does not need to be a complex process. Training may be as simple as sitting around the table discussing standard operating procedures (SOP) or as complex as a field exercise that involves multiple agencies. One of the easiest ways to train on the application of information in the field is through exercises. An exercise is "a focused practice activity that places participants in a simulated situation requiring them to function in the capacity that would be expected of them in a real event."[1] Not only do exercises place you in a simulated scenario, but they also allow for the evaluation of your actions as well as evaluation of any policies and procedures that are utilized during the exercise. Exercises, when conducted in conjunction with other agencies, allow you and your agency to interact and strengthen response relationships with other public and private entities.

Exercises are also important to conduct because they may meet regulatory requirements that are mandated by local, state, and federal agencies. If you have a nuclear power plant in your jurisdiction, you may want to coordinate an exercise with their personnel, as they are required to conduct a full-scale exercise every two years. Airports, hospitals, and other health care facilities also face requirements of exercises, normally every two years. Your own agency may have a requirement for emergency plan activation based on OSHA requirements. Even if you do not have a facility that is required to exercise emergency response plans with your response area, you can still work to conduct an exercise that benefits your agency and assists others in preparing for emergency response.

Exercises allow agencies to:

- Test and evaluate plans, policies, and procedures.

- Reveal planning weaknesses.

- Reveal gaps in resources.

- Improve organizational coordination and communications.

- Clarify roles and responsibilities.

- Train personnel in roles and responsibilities.

- Improve individual performance.

- Gain program recognition and support of officials.

- Satisfy regulatory requirements.

Scenario: After running through whiteboard scenarios with your shift, you realize that there is a need to review the ICS structure from the very basic level. The members of the shift definitely seem to need an opportunity to review their knowledge, skills, and abilities as they apply to ICS. In asking around the shift and even other shifts, you realize that everyone seems comfortable with the everyday "simple" incidents. Everyone admits that they have a weakness when it comes to setting up an ICS structure in a more complex incident, and they are least familiar with complex incidents involving mutual aid resources. You approach the captain and ask to speak to him.

"Captain, is there a way we can do more training on setting up an ICS structure at a more complex incident? We understand the everyday, single car accident structure, but everyone agrees that a larger incident, like an MCI, is something we would not be comfortable with."

The captain agrees. He puts you in charge of organizing a training event for the station. He tells you to do the research and determine what type of exercise is best for the rest of the shift. He reminds you that he will expect a report at the completion of the exercise to determine what changes, if any, need to occur to improve shift efficiency and effectiveness.

Where do you begin? It cannot be as easy as it seems. There has to be more to developing an exercise than writing an exercise, bringing everybody into a room, and then getting them to talk. Where do you start? What did he mean by "type of exercise"? How many different types can there be? You realize that maybe you have gotten yourself into more than you realized, but get started because you want to develop something that truly benefits the shift and the citizens.

Exercise Development

In recognizing the need for review of information, you can take it upon yourself to prepare exercises for you, your crew, your agency, and other agencies you work with. While it takes more than a single person to develop a full-scale exercise, you can begin the process of determining which training exercise meet your agency's needs.

The first step is conducting a needs assessment. This assessment should begin at a small scale, with the provider in the field. In the scenario above, you, your partner, and other personnel assigned to your station should determine what you feel you need in the way of exercise. A needs assessment assists you in recognizing the high probability risks that you have within your jurisdiction. Once you have determined the risks, you should prioritize them based on the potential of occurrence with your jurisdiction or jurisdictions to which you provide mutual aid. The highest priorities should become the focus of upcoming exercise scenarios. You can also determine if your locality has already conducted a hazard analysis, which will look at the incidents that have the greatest potential of occurrence. This can assist you in determining exercise needs. You can use a needs assessment for the station to design and conduct station level exercises. You can also provide your needs assessment to your superiors in the hopes that it will be utilized in an agency-wide exercise.

Once you have conducted a needs assessment, you need to determine what your desired outcome is after conducting an exercise. For example, do you want to determine resource needs? Do you want to determine training needs? Do you want to determine if a specific SOP addresses the appropriate issues? Each of these goals affect how the exercise is conducted. The desired outcome will influence the structure, participants, and even the type of exercise conducted.

Deciding on an exercise type

The Department of Homeland Security has developed a Homeland Security Exercise and Evaluation Program (HSEEP).[2] The HSEEP provides a method for developing and evaluating an exercise program within an agency. The program recognizes seven different types of exercises. Four of the exercises are considered "discussion-based" exercises. The discussion-based exercises focus on verbal discussion or classroom practical evolutions instead of full hands-on activities. These types of exercises provide an opportunity for you and fellow participants to become familiar with the policies, plans, and procedures of your agency and locality as it pertains to the exercise scenario. Discussion-based exercises focus on the use of personnel, not resources, to accomplish exercise goals, meaning that they normally require less preparation and preplanning. The four discussion-based exercises are:

1. Seminar. A seminar provides an informal environment in which discussions regarding new and old policies and procedures orient providers to these documents.

2. Workshop. Like a seminar, a workshop provides an opportunity for informal discussions. However, a workshop is designed to provide an opportunity for the development of a product such as a plan or policy.

3. Tabletop exercise (TTX). A TTX brings together leadership and other key personnel to work through scenarios in an informal setting without full, hands-on activities. Many of the actions that would be taken are verbalized, with the TTX providing only a visualization of the scenario.

4. Game. A game is a simulation that places two teams against each other while running a scenario based on rules and policies. Games are designed to be a competition against the two participants.

The second group of exercises are operational-based exercises. These exercises serve as a method of validating plans, policies, agreements, and procedures that are in place. These documents may be validated, proving that they accomplish their goals, or it may be determined based on the exercise that they need to be modified to better fit the needs of your agency. An operational-based exercise can also provide an opportunity to clarify roles and responsibilities of the responders and leadership and may provide an opportunity to recognize operational deficiencies. Operational-based exercises include:

1. Drill. A drill is a hands-on, coordinated activity normally used to test only one specific aspect or function. A drill for EMS providers might focus on the triage capabilities of the agency.

2. Functional exercise (FE). A functional exercise provides the opportunity to exercise the coordination, command, and control between response agencies. Functional exercises

involve the leadership instead of those who actively partici-pate in the field operations of an incident. An FE provides the opportunity for leadership of multiple agencies to partic-ipate in a review and determine what policies may need to be strengthened or changes need to be made.

3. Full-scale exercise (FSE). A full-scale exercise involves responders of all levels, from the field personnel to those who serve in the multi-agency coordination centers (e.g., emergency operation centers, joint field offices, etc.). An FSE provides an opportunity to work through a simulated scenario, while actually conducting the activities that would happen in a real-life situation.

Table 6–1 provides a side-by-side overview of the goals of each type of exercise. Each of these exercises provides an opportunity to practice and learn. The operational exercises take a greater amount of money and time to plan and conduct, but may provide greater benefits, depending on the purpose of the exercise.

Table 6–1 Overview of exercise goals

Seminar	Familiarize participants with new or existing guidance documents. Research or assess interagency capabilities or jurisdictional operations. Construct a common framework of understanding.
Workshop	Develop new ideas, processes, or procedures. Develop a written product as a group in coordinated activities. Obtain consensus. Collect or share information.
Tabletop	Identify strengths and shortfalls. Enhance understanding of new concepts. Seek the change of current attitude and perception.
Drill	Become familiar with new equipment. Test procedures. Practice skills.
Functional	Evaluate command structure functionality (EOC, ICP, etc.). Determine adequacy of resources, response plans, and equipment.
Full-Scale	Assess response plans and procedures in real-time scenarios. Evaluate multiagency response coordination under emergency situations.

Designing the exercise

Once you have chosen the appropriate type of exercise, the process is the same. The next step, whether you have chosen a discussion-based or operational-based exercise, is to begin the exercise design process. The first step has already been completed: Complete a needs assessment. This needs assessment provides guidance in determining the type of exercise to conduct, and will also assist in the design process. Once the needs assessment is complete, you need to define the scope. Without a well-defined exercise scope, your exercise can quickly lose focus and become disorganized. By defining the scope of your exercise you are looking realistically at the goals that can be accomplished. This assists in determining the participants and also the resources needed. By defining a scope you are also defining the priorities of your exercise. Many factors can affect your exercise scope. Some of these include:

- Expense. If money is not available to support overtime and other aspects of the exercise (e.g., logistical support, equipment, supplies, etc.), then the exercise will need to be on a smaller scale and involve fewer people.

- Availability of personnel and other resources. If trucks and people are unavailable, then it is impossible to hold a full-scale exercise. In the case where large numbers of personnel are available, a full-scale exercise may be the most beneficial scenario.

- Length of the exercise. If you determine you only have a small time frame in which to conduct the exercise, simple is better. The more hands-on an exercise is designed to be, the longer it will take to conduct.

There are five key elements to an exercise scope that must be defined during the development process. The first is the type of emergency. Based on the needs assessment, you need to determine what type of emergency incident you will exercise. The focus should be those that have the highest potential to occur in your locality, those that have not been exercised recently, or those problems that may have just developed for your department. Choosing the type of incident also helps you determine the type of exercise, the second element in the exercise scope. Once you have determined the type of incident and type of exercise to be conducted, you need to determine where the exercise will be held, the third element. The type of exercise will help in determining the space needed. The space should make the exercise as realistic as possible. If possible, the exercise should be conducted at the facility or in an area where the event may occur. The fourth key element in defining scope is determining the functions that will be conducted. The functions of the exercise should be clear and concise to allow for minimal confusion of event participants. With functions determined, you can focus on the final element of the scope of the exercise: the participants. In a functional exercise, the participants focus more on the leadership of the participating agencies. In a workshop, the participants may be a set of first responders from a set group of stations. After developing the scope of the exercise, you need to determine a purpose. When written, the statement of purpose should not be too specific, ensuring that it applies to the whole exercise. A purpose statement helps determine and clarify objectives and participants and provides a method of providing information about the exercise to community leadership.

Step four of exercise design is development of objectives. Objectives provide a desired action for the participants to meet. Objectives should be written for all stages of the exercise, from design to evaluation. Exercise objectives should be written based on the exercise itself. If the exercise involves a small number of participants, only one or two objectives may be necessary. If the incident is large, such as a full-scale exercise, then a significant number of objectives may be necessary. In exercises that involve multiple agencies, each agency should be responsible for developing its own objectives, which then become integrated into one exercise plan. This ensures that objectives are written by those who understand the roles and responsibilities of the agencies involved.

WRITING A SMART OBJECTIVE

Simple   The objective should be clearly phrased and
                    easy to understand.

Measurable   The objective should set a performance goal
                    that is measurable and can be observed.

Achievable   The objective should not be too difficult to
                    achieve. Do not set exercise participants up to
                    fail in achieving the objectives.

Realistic   The objective should be written on realistic
                    expectations for the given situation.

Task oriented   The objective should focus on a specific behavior
                    or procedure.

Step five in the exercise design process is when you get to start writing the story or narrative. The narrative is the opportunity to provide a vivid story for the participants of the exercise and begin to develop the initial response activities. The narrative should provide a clear and concise story that motivates the participants to actively participate in the exercise. When the exercise focuses on an event that provides some lead time to prepare for response, such as a hurricane, the narrative should include a well developed outline of the activities leading up to the timeframe the exercise is set in. For events that provide no preparation time, such as a large highway incident or a terrorist attack, the narrative may include a brief description of the day and the initial notification information. Remember though, that as much as you may want to make the narrative detailed, it should provide only the initial information for the event.

EXAMPLE NARRATIVE—UNKNOWN EVENT

It is 5:30 on a sunny Saturday afternoon. The temperature is 92° F with a slight breeze from the east at 5 mph. Humidity is 72%. The Jackson City speedway is preparing for the last race of the season. Attendees are starting to enter and take their seats in the stands. The drivers complete their final preparations and have lined their cars up for the start of the race. The attendees all rise to their feet for the singing of the national anthem and the order for the drivers to start their engines.

As the crowd claps at the completion of the national anthem, the stands collapse, taking approximately 5,000 attendees to the ground.

The speedway's on-site ambulance responds first, calling the local 911 center to request additional resources to assist in the technical rescue, triage, treatment, and transport.

EXAMPLE NARRATIVE—KNOWN EVENT

While watching the nightly news, you are inundated with information on the incoming hurricane. The three-day forecast of hurricane Catherine places landfall just north of your county. The National Weather Service (NWS) has issued a hurricane watch for a three-state area along the coastline. Winds of 135 mph are estimated at landfall and the water is projected to be 15–20 feet above normal high tide. Access to bridges is narrow and could be easily blocked by high water. Downtown has known drainage problems, so significant flooding is expected.

Evacuation of the coastal areas has begun with approximately 15,000 residents participating, meaning that the main highway access to the coastal areas is no longer passable. Inland shelters are open for those who are unable to evacuate.

State emergency management personnel have notified the localities that the state emergency operations center (EOC) will be augmenting staffing at 0700 hours tomorrow morning. All coastal localities have already staffed their EOCs and emergency response personnel have begun emergency staffing procedures. The local emergency manager has called a meeting for emergency services personnel at 0730 tomorrow morning to discuss preparedness and response activities within the locality.

With a narrative written, it is time to work steps six and seven of exercise design. These two steps may be completed concurrently as they have a direct connection to each other. Step six of exercise design is writing the detailed events and step seven is determining and listing desired actions. Detailed events push for a specific action to occur within the exercise. Desired actions are the specific behaviors or decisions that you want participants to complete during the exercise. Desired events should signal the beginning of a specific problem and should provide specific information about the issue. The events should be written with the desired outcome in mind. Because of this, you may choose to complete step seven, desired actions, before you complete step six. You may also choose to work the two steps at the same time.

Consider the sample narrative referring to the speedway incident. Table 6–2 provides a selection of sample events and sample actions. As you are writing the events and actions, keep in mind the predetermined objectives of the exercise. You want to ensure that the events and actions direct the participants toward the objectives. For exercises that involve a large number of agencies with different objectives, involve a representative from each agency to design the events and actions. Also, remember that multiple events can occur concurrently to ensure that all participants have a function.

Table 6–2 Sample of desired events and desired actions

Desired Event	Desired Action
Upon arrival on scene you see a large area of unsecured stands and attendees trapped within the wreckage. Those who were not involved are trying to get to those who are injured.	Establish scene control and Incident Command Structure to begin good incident response management.
Firefighters have secured a small portion of the stands and you now have access to a limited number of victims.	Establish triage procedures and set up a treatment area to receive patients, once triaged.
Your triage team has completed their tasks and reports back that they have the following numbers: 60 Black; 42 Red; 50 Yellow; 150 Green.	Request appropriate additional transport units upon recognition of resource needs.

Exercise messaging

Once the list of events and actions is complete, you must determine how the participants will be notified of each event during the exercise. To keep the exercise as realistic as possible, remember that at a real event messages may be sent through a variety of ways, including:

- Radio

- Cell phone

- Landline

- In person

- Text messaging

- Fax

- Written documentation

SAMPLE EXERCISE MESSAGES—SPEEDWAY INCIDENT

Message via radio

FROM: Technical rescue team   **TO: Operations section chief**

Operations section chief from technical rescue team. Be advised we have stabilized the first section of bleachers and it is now safe for the triage team to enter.

Message via radio

FROM: Triage team leader   **TO: Operations section chief**

Operations section chief from triage. Be advised triage is complete and we have the following numbers: 60 black, 42 red, 50 yellow, 150 green.

Messages should be written as if they were being transmitted in real time at a real event and must contain credible information. They must also be tied to an expected event and action to ensure that the exercise is progressing as designed. Consider all the aspects of the message during the development phase. First determine who is sending the message. Make sure that the message is written as if it was from a source that would send the message. The transportation officer would not call the incident commander with triage numbers; that would be sent from the triage officer. The next thing to consider is how the message would be transmitted. If in a real life scenario a radio would be used, then during your exercise ensure message transmission is made over a radio. After you have determined the message content and transmission method, consider who the message is being sent to. This is determined by the actions you want accomplished. A message designed to initiate actions toward a communication relationship between the hospitals and the treatment area would not be sent to the triage officer, it would be sent to transportation. Remember that everything does not always go right during an exercise. If you find that the participants are straying from the intended actions, messages may need to be scripted on the fly. Be prepared to improvise messages during the event to ensure that the exercise stays on course.

Specific Exercise Planning

Prior to determining which type of exercise to utilize, the planning process is consistent, allowing for a generic approach to determining objectives and exercise goals. However, differences exist in the final steps of exercise development and implementation. Until this point, this chapter has focused on the generic development aspects. The next portion of the chapter looks at specific exercises and specifics in their development and implementation.

Tabletop exercise (TTX) considerations

As previously mentioned, a tabletop exercise provides an opportunity for an informal review of the processes associated with a specific incident. Because of the significant number of advantages offered through the utilization of a tabletop, this method is often selected for exercises over a more hands-on approach. Advantages of tabletop exercises include efficiency in cost, time, and resources; ability to integrate any situation into the scenario; and they are a good and safe way for key personnel to be introduced and review responsibilities and procedures of their own department as well as others. While there are disadvantages to the use of a tabletop exercise, such as the limitations of reality of the incident, the ease of conducting a TTX makes it appealing to many.

One of the issues specific to each exercise is the method of delivering exercise messages. While a tabletop exercise provides little opportunity to present a realistic delivery method, there are options. The first option is to use family service radios. Family service radios are portable two-way radios that allow for traffic on different channels. They not only simulate radio communication and can assist in testing communication plans, but they also provide an opportunity to simulate the chaos that can occur with uncontrolled radio traffic during an incident. In most uses, face-to-face verbal delivery of exercises messages is the method utilized during tabletop exercises. While easy to control, realism is easily lost, so the *who* and *what* of message delivery is key. If the message would be given to a large group to make decisions, then the entire group of participants should receive the message. If, in a real situation, one person would receive the message, that is how the tabletop should proceed. Remember, sometimes the reality of the situation proves to be the best lesson learned. In a TTX, once the messages are delivered, instead of physically acting on the message, participants discuss the actions they might take in a real incident. Another approach to the TTX is to provide a message to the desired recipient, allowing that person to make a decision on the actions to take as a result of the message, and then allow all of the participants to discuss the individual's decision. This allows for a great discussion base and may bring up suggestions and ideas for improvement.

Sustaining activity of a tabletop may prove a challenge for the facilitator. Ensuring that members stay on task and focus on the purpose of the tabletop takes guided effort. It is easy for participants to lose focus and begin discussions that branch away from the main goal of the exercise. Because of that, it is the facilitator's responsibility to maintain control and sustain activity. Having multiple messages to assist in feeding the participants with exercise related information helps to maintain their focus. That way, if discussion fades on one topic, you can introduce a new topic for discussion. Another method of maintaining activity and discussion is to be flexible. Flexibility allows you to delete a message when the pace is moving well or to add additional messages when the pace is slowing down. However, you also want to make sure that you provide an opportunity to appropriately discuss all items. Make sure that too much time is not given to one message and not enough time given to another.

Another thing to remember while facilitating a tabletop is that conflicts may arise. Unlike field exercises, participants in a tabletop maintain close proximity to each other, meaning that there is little break from the energy of the exercise. Because some participants may feel strongly about their policies and procedures, discussions that seem critical may lead to defensive attitudes. The tabletop facilitator must ensure that first and foremost, discussion is beneficial, not critical of any participant. If this does occur, the facilitator should recognize the tension, stop the exercise, and deal with any issues before moving forward. This ensures that the tabletop is continued in a non-hostile environment and in a productive manner.

In terms of the design of the exercise, not much is different. An alternative to the standard tabletop does exist that allows jurisdictions to exercise within their own ranks as well as to interact with neighboring jurisdictions. In this type of exercise, neighboring jurisdictions conduct a TTX concurrently. They work the local aspect of the same scenario within their tabletop, but feed information to the other jurisdictions through fax, e-mail, telephone, and radio communications. This allows participants to exercise their own policies and procedures while determining strengths and weaknesses of the communications system between neighboring localities, thus strengthening the overall exercise.

Regardless of the delivery or method you use for a tabletop exercise, always keep in mind the objectives you are trying to achieve.

Functional exercise considerations

As previously stated, a functional exercise is one in which participants test a specific function of incident response. While every type of exercise evaluates certain functions, a functional exercise has key characteristics that differentiate it from other types of exercises.

A functional exercise brings together a large number of people, even if from only one jurisdiction. As with all exercises, bringing together the right people is necessary in ensuring a successful event. The players in a functional exercise represent the individuals who hold leadership or coordinating responsibilities. Because a functional exercise tests the plans, policies, and procedures of an organization, it is important that the people who make the changes to those documents play a vital role in the exercise. It is the responsibility of the participants to respond to the simulation as they would in a real-life scenario, not as they believe the evaluators want them to respond. If the exercise does not include the function the participant would normally do, then that individual does not need to have an active role in the exercise.

KEY CHARACTERISTICS OF A FUNCTIONAL EXERCISE

- Requires significant scripting and planning due to the length and complexity

- Involves higher level personnel and normally takes place in the EOC or an operations center

- Is normally geared for policy, coordination, and operational personnel

- Actions and messages are given in real time and are designed to generate real-time responses and consequences

- Atmosphere is stressful and tense as the result of the realism and message response time requirements to keep the exercise flowing

Unlike a tabletop exercise, simulators are required to assist in the implementation of the exercise. In a tabletop exercise, the coordinator serves as the simulator, facilitator, and evaluator. A functional exercise requires the use of separate simulators. Simulators serve the role of message senders, and should be able to change the messages as the exercise dictates. Individuals filling the role of simulator should be familiar with the exercise and the expected actions of the participants. This allows them to recognize when the actions are not going to be as expected, which gives them time to prepare a message response to redirect the exercise back to the objectives. The number of simulators needed for an exercise is dependent on a number of factors, including the size of the exercise. If the exercise involves a large number of players, then multiple simulators are necessary to ensure that all players are active throughout the exercise. Additionally, the length of the exercise can impact the number of simulators involved. As a result of the constant work of simulators, this position is best staffed in short shifts, allowing for breaks during the exercise. Regardless of the number of simulators, make sure that the individuals chosen are familiar with the expected participants, their functions, and their potential actions.

A second player that is introduced in a functional exercise is the controller. The controller serves as the overall exercise manager. The evaluator ensures that he or she maintains a view of the whole exercise, not just one part, as a way to manage all personnel. Responsibilities of the controller include pre-exercise training of evaluators and simulators, introducing participants to the exercise, overseeing adjustments to the scenario as dictated by exercise play, adjusting the pace of the exercise, and most importantly ensuring order and professionalism is maintained throughout the exercise. As with the simulators, the controller of a functional exercise requires specialized knowledge, skills, and abilities. The individual chosen to fill the role of controller must be able to multitask, managing both the participants and the exercise.

Once the players of the exercise are chosen and the exercise is developed, the functional exercise can begin. A functional exercise is developed like other exercises. The desired function is chosen, a list of events and actions is developed, and messages are created from this list. The timing of the exercise is then determined by the exercise objectives. If the exercise is to determine after-hours notification procedures, then participants may only be given a general time frame to expect activation. This provides a more realistic aspect to the exercise. In an exercise that is not based on immediate response actions, notification is conducted in advance. When the participants gather for the exercise, they are briefed with an overview of the expectations, guidelines, and procedures of the exercise. This provides an understanding of the day's events. It also allows you to ensure that the focus is placed on the exercise and not activities happening outside of the exercise. Distractions from the outside environment should be minimized, if not eliminated. Once the introductions are complete and the initial narrative is read, the exercise begins with simulators providing messages and feedback based on participants' actions. Even with guided messages, remember that you may need to encourage the participants to act as if this is a real incident. Because the exercise is not set in the same environment as a real incident, it may take some encouragement to get them to act as if this were a real event.

Supporting a functional exercise requires not only the exercise messages and personnel, but equipment and locations that provide sufficient space and a realistic setting. If the event is to exercise the functionality of the local EOC activation procedures, then participants should gather at the local EOC. The supplies needed to support it should also mirror what may be on hand during a real event. Dry erase boards, easels, computers, and communications equipment may all be necessary to help facilitate the exercise. This equipment supports participants, evaluators, and simulators throughout the exercise. Determine resource needs before hosting the exercise so the necessary equipment can be secured. For the simulators, additional radio, phone, and electronic communications methods may need to be installed to support message sending. All of this should occur prior to conducting the exercise.

Full-scale exercise considerations

A full-scale exercise is a huge undertaking, but provides an exercise that simulates a real event better than any other method. It allows for real time, real action, real patient scenarios that give realistic practice for participants and provide an opportunity to get the most realistic review of policies, plans, and procedures. Full-scale exercises are a requirement of many facilities including nuclear power plants, airports, and hospitals. As such, you and your agency should work with facilities in your area that are required to conduct full-scale exercises. You can also choose to conduct your own full-scale exercise. In order to receive credit from FEMA, the exercise must meet three requirements:

1. It must exercise most functions that are described by FEMA.

2. It must coordinate the effort of several agencies in response to the exercise scenario.

3. EOC activation must be included in the exercise to achieve full coordination of participating agencies.

Based on these requirements, planning a full-scale exercise takes significantly more work than any other type of exercise.

Full-scale exercises display some specific and unique characteristics and some that are shared with other exercises.

KEY CHARACTERISTICS OF A FULL-SCALE EXERCISE[3]

• Has an interactive design geared to involve all agency members, from operations to leadership, and provides a realistic situation in a stressful environment

• Provides the opportunity to evaluate a majority of the functions of an emergency management or operational plan

• Takes place at multiple locations including an EOC and field sites, normally across multiple jurisdictions

• Achieves realism through a variety of methods, including:

— Simulated "patients/victims"
— Search and rescue activities

— Use of communications equipment

— Resource and personnel allocation as necessary to work the scenario

— Deployment and use of field equipment

— On-scene activities, decisions, and related consequences

- Involves a controller, players, simulators (different from those used in a functional exercise), evaluators, and observers

- Includes players representing all levels of a participating jurisdictions, from response personnel to management and executive level staff

- Provides messages for individual events (e.g., staged incidents, moulaged patients, etc.)

- Allows decisions to be made by exercise participants in a real-time environment with real responses and consequences resulting

- Requires significant investment of time, money, effort, and resources in the development and implementation

The development of a full-scale exercise requires a mix of patience, time, and assistance to ensure that all details are handled and a successful exercise occurs. The planning team for a full-scale exercise is normally very large, with representatives from the local to federal level participating in the development of the objectives and scenarios. Because gathering the logistical support ranges from securing appropriate facilities to securing and putting into place appropriate equipment, planning of a full-scale exercise can take from one to one and a half years.

Once you have gained support for the time, money, and resources it takes to conduct a full-scale exercise, you can move to the planning process. As with any exercise, the players, from those who are facilitating to those who are participating, need to be determined. The following provides an overview of the players in a full-scale exercise:

- Participants. A full-scale exercise requires participation from all levels of the organization. Policy makers, support personnel, operations personnel, and field responders all play a role in the exercise.

- Simulators. Different from the simulators in a functional exercise, these simulators serve as "victims." Their roles are scripted to simulate injuries or other issues and makeup may be used to further support injury simulation.

- Evaluators. Evaluators of a full-scale exercise fill the same role as evaluators in other exercises. Because of the expanded area of a full-scale exercise and the multitude of scenarios being run, the number of evaluators is significantly greater in a full-scale exercise.

- Safety officer. As with any real incident, a full-scale exercise must include a safety officer. The safety officer is staffed during the planning process and reviews the exercise to determine safety issues and concerns. The safety officer makes recommendations to improve the safety of the exercise.

Each of these people plays an important role in the development and implementation of the exercise.

Once the players are chosen, focus can switch to the exercise development. The steps for the development of the exercise are the same, but the details can become overwhelming. Remember that a full-scale exercise requires intensive commitment of personnel and resources. Both private and public agencies must recognize the potential overtime and financial implications of participation in a full-scale exercise. Not only do jurisdictions need to consider overtime, but the fuel and operating costs of resources will also impact participation decisions. Funding may be available through grant opportunities through the state and federal agencies.

With the players chosen and objectives and messages written, the full-scale exercise begins. Because of the reality of the situation, people respond to messages as they would in a real situation. If the role requires reporting to the EOC, then that is where they go. If the role is serving as ALS providers on an ambulance for a shift, then they report to the station for their shift and respond as the message dictates. Because exercise activities cover a large geographic area, a centralized "dispatch" center can be utilized to coordinate exercise messages. This dispatch center serves as the central point of message dissemination, especially those sent and received via radio communications.

Scenario: You and the committee have finished the development process for the exercise. It was simple enough until you are reminded of the fact that you cannot serve as both the facilitator and the exercise evaluator. At the next planning meeting you call everyone to order and post to them the next step, evaluation development. With that, the questions start quickly. With all of these questions you cannot help but wonder where you start.

Who will be tapped to evaluate the exercise? What makes a good evaluator? What specifics will the evaluators look at/for? What can be done to ensure that they are watching for the appropriate things and providing feedback in a standard manner so it can be reviewed and utilized to improve performance of the providers in the field?

Evaluating the Exercise

Development of your exercise evaluation tools can occur at the same time you are developing the exercise itself. Knowing what you hope to gain from the exercise (the exercise goals) is the first step in working toward a successful evaluation tool. Remember, an exercise is important for the system, but even more important is the lessons the exercise provides. In order to capitalize on the opportunity to learn from an exercise, you have to develop an evaluation tool that allows you to recognize the strengths and weaknesses of the actions of the responders, whether from preset policies and procedures or from the actions of the participants during the exercise. Evaluation can also assist you in recognizing the strengths and weaknesses of the exercise itself, allowing you to improve future exercises. The HSEEP program offered through the Department of Homeland Security provides guidance in developing an after action review (AAR) process and format and also provides electronic documentation to support the development and implementation of a corrective action plan (CAP), a concept that will discussed later in this chapter.

Developing an evaluation and improvement program

As previously stated, the HSEEP program provides a process for the development of an evaluation and improvement program. As with the development of an exercise, the development process for evaluation occurs through a process of steps to ensure that all needs are met. The exercise development team also completes the evaluation development. The evaluation stage occurs throughout the entire process, from prior to the actual exercise through the completion of the exercise and after. The first step of the evaluation development process is planning and organization.[4,5] For larger exercises, evaluation planning is best conducted by a committee so that each participating agency can have representation. Like the exercise design committee, the evaluation planning committee holds responsibility for development of an evaluation tool, choosing the evaluators, and preparing for the evaluation process.

Evaluation team. The size of the exercise dictates the size of the team. Full-scale exercises may require an overall evaluation group director with multiple team leaders to support evaluation of scenarios at multiple sites, while a tabletop exercise may only require a single evaluator. Once you determine how many evaluators you want, you must determine who will fill the evaluation role. When selecting evaluators keep in mind their knowledge, skills, and abilities. These are as important as their knowledge of the function they are evaluating. Table 6–3 lists the suggested evaluator traits. These characteristics allow evaluators to be successful in their role and provide useful information to you and your agency.

Table 6–3 Suggested traits of an exercise evaluator

Knowledge	Technical expertise in evaluation Understanding of function Understanding of relationship between desired actions and objectives
Skills	Communication skills, both verbal and written Organizational ability Ability to adjust to rapidly changing situations
Attributes	"People skills" Objectivity Willingness to help Honesty and integrity (reports facts truthfully, keeps information confidential) Familiarity with the exercise plan

EXAMINING EXERCISES

In order to assist in deciding which information will be evaluated, the Department of Homeland Security (DHS) provides a suggested three-level approach to evaluating exercise performance.[6]

Level one

Level one is the task level. The task level focuses on evaluating the individual's performance of required tasks throughout the incident.

This part of the evaluation may look at specific tasks each person attempts to accomplish during the exercise.

Level two

Level two is the organizational/discipline/function level. When evaluating tasks at this level you focus on the ability of an organization or agency to complete tasks that are specific to their functions (e.g., hazmat teams conducting hazmat response, EMS agencies providing emergency medical care, etc.).

Level three

Level three is the mission level. When evaluating at this level, you place focus on the ability of all agencies to work together to complete the incident objectives.

Remember that in a complex exercise that not every level will be evaluated, because not every level is exercised.

Evaluating the exercise

When determining which information will be collected, remember to keep the exercise goals in mind. Information should assist in reaching the goals and determining if the policies and procedures are effective and efficient in reaching the goals. With clearly stated objectives, the information that needs to be gathered has already been determined, it is the method of the evaluation that the planning committee needs to consider. Job aids play an important role in assisting evaluators in gathering data from the exercise. Checklists and guided steps can assist evaluators in making sure that they are reviewing and monitoring the appropriate actions and gathering the desired information. Forms also allow for a uniform structure for gathering the information, which ensures that the review of the evaluations is easier to conduct.

One type of monitoring that should be conducted is key event monitoring. When main messages are sent, evaluators monitor the participants affected by the message to document their actions. These actions may be evaluated to determine the effectiveness of the policy, the participants' ability to follow policies, or the need to improve policies and procedures. Additional monitoring may be conducted on the leadership, and even on the messages themselves to determine the effectiveness of the exercise.

Along with the visual assessment conducted by the exercise evaluators, a real-time evaluation is also completed by all participants. A problem log provides an opportunity for participants, controllers, facilitators, and simulators to document any problems they see during the exercise. One of the things to remember in reviewing the problem log is that what may have been perceived as a problem during the exercise was actually a simulation designed to elicit specific actions from the participants. Another evaluation tool that may be used is an exercise critique form. This form allows individuals participating in the exercise to review the exercise format and design.

Step two of evaluation occurs during the actual exercise. This is when the evaluators actively watch and record data during the exercise and the participants. Information should be recorded in the predetermined format for easy analysis at the completion of the exercise. Not only should evaluators utilize the actual actions and words of the participants, but they should also review any written logs or records to conduct their evaluations. With the data from the exercise collected, you can move to step three.

Step three of the evaluation and improvement program is the data analysis. The information gathered is summarized and reviewed for numerous purposes. One of the first is to determine if the expected action matches the actual performance. This may lead to changes in training on policies and procedures to ensure that agency members are familiar with them. Evaluations can also be reviewed to determine if any best practices were displayed that can be put into policy.

With data analysis complete, step four begins: the development of a draft after action report (AAR). An AAR is a summary of the exercise and provides an overview of the information gathered from the data analysis, highlights the issues that were recognized, and provides recommendations for improvements. This report, developed by the evaluation team leader and evaluation team, follows no specific format, though table 6–4 provides a list and description of elements that should be contained within an AAR. The appendix contains a template for an after action report as designed by FEMA for the Homeland Security Exercise and Evaluation Program (HSEEP).

Table 6–4 Minimum data elements in an AAR

Data Elements in an After Action Report
Executive Summary
Exercise Overview—Includes purpose of the exercise, participating agencies and jurisdictions, exercise information (date, time, place, type, etc.)
Exercise Goals and Objectives
Scenario Overview
Chronological synopsis of major events and actions.
Analysis of Mission Outcomes
Summarizes how the performance or nonperformance of tasks and interactions affected achievement of the mission outcomes.
Analysis of Critical Task Performance
Summarizes and addresses issues regarding each task in terms of consequences, analysis, recommendations, and improvement actions.
Conclusion
Appendix: Improvement Plan Matrix
Provides a task list of recommendations, due dates, and responsible organization

The AAR provides an opportunity to both praise and provide critical review of the actions of the participants as well as the guidance documents of each participating agency. It allows for the sharing of information with all participants to ensure that they receive a complete report of their activities, not just ones developed by their individual evaluators. The AAR also allows for a review of interagency coordination and a chance to provide feedback on improving emergency response as a single agency and as a whole jurisdiction. Once complete, the draft AAR is shared with all participating agencies and jurisdictions. At this point an after action conference occurs, which is step five of the evaluation process. The after action conference provides an opportunity for the evaluation team to present their AAR drafts to the leaders and key members of participating agencies. It also allows for open discussion and feedback regarding the information shared.

Step six of the evaluation process is when the recommendations are reviewed and decisions are made about what changes people believe would best benefit their agencies. During this step improvements to be implemented are identified by participants. Each participating agency may have a different view of which changes and recommendations are best, but any recommendation is written in the AAR document. Once improvements are identified, the steps necessary to implement the improvements are identified and a suggested time frame for implementation is developed. This allows for a structure to the change, so that participants do not freelance with the changes they feel are best for their organization. With this in mind, agencies also need to appoint an individual to supervise the implementation process and oversee the changes. If multiple changes need to be implemented, a separate time line for each item should be created. This ensures that each change can be efficiently monitored to determine its effectiveness.

With the implementation guide created, the final AAR is produced and distributed to all personnel. This is step seven. The final after action report should include information from all steps of the event, from development to analysis. Such a comprehensive report will provide the most information and best benefit to those who review it.

At the completion of the evaluation process is the review of changes that have been implemented to ensure efficiency, effectiveness, and appropriateness within each agency (step eight). Review of the benefits of the changes is done by the agency/jurisdiction in recognition of all of the roles within the agency. Change should never be made without an opportunity to review the impact the change has on the people and the agency's function. After a set period that is long enough to allow data collection regarding the changes, the designated individual(s) should review the new procedures and determine whether they still make sense, if the agency has the resources to sustain and support the changes, and if the personnel impacted by the changes are appropriately trained and knowledgeable about the changes and the reasoning behind them. If the review finds that a change is not beneficial, then the agency should suspend the change until further review can be done or additional suggestions are made for system improvement.

Conclusion

Training is a continual process that provides new information and an opportunity to review previously learned information to the participants. Multiple learning methodologies exist including classroom, lecture, hands-on, online, and hybrid or combinations of multiple training methods. In the emergency response field, hands-on training is often the best method for reinforcing the knowledge provided in initial certification courses, especially when focusing on incident command training.

References

1. U.S. Department of Homeland Security, Federal Emergency Management Agency (FEMA) (March 2003). "IS-139: Exercise Design," in Emergency Management Institute web site. Retrieved June 1, 2011, from http://training.fema.gov/EMIWeb/IS/is139.asp.

2. U.S. Department of Homeland Security Homeland Security Exercise and Evaluation Program (HSEEP) (February 2007). *Volume I: HSEEP Overview and Exercise Program Management*. Retrieved September 1, 2010, from https://hseep.dhs.gov/support/VolumeI.pdf.

3. FEMA (January 16, 2008). *IS-130: Exercise Evaluation and Improvement Planning.* "Lesson 3: Planning and Organizing the Evaluation," in Emergency Management Institute web site. Retrieved June 1, 2011, from http://emilms.fema.gov/IS130/EXEV0103000.htm.

4. See note 3 above.

5. FEMA (March 3, 2010). *IS-120.a: An Introduction to Exercises*, in Emergency Management Institute web site. Retrieved September 1, 2010, from http://training.fema.gov/EMIWeb/IS/IS120a.asp.

6. FEMA (January 16, 2008). *IS-130: Exercise Evaluation and Improvement Planning.* "Lesson 6: The After-Action Report and After-Action Conference," in Emergency Management Institute web site. Retrieved September 1, 2010 from http://emilms.fema.gov/IS130/EXEV0103000.htm.

[Protective Marking]
Homeland Security Exercise and Evaluation Program (HSEEP)

After Action Report/Improvement Plan	**[Full Exercise Name]**
(AAR/IP)	**[Exercise Name Continued]**

[Note for after action report/improvement plan (AAR/IP) template:

- Text found in this document that is highlighted and bracketed is included to provide instruction or to indicate a location to input text.

- All text that is not highlighted is to be included in the final version of the AAR/IP.]

[Full Exercise Name]
[Exercise dates]

AFTER ACTION REPORT/IMPROVEMENT PLAN

[Publication date]

[On the cover page, insert additional graphics such as logos, pictures, and background colors as desired. The word "Draft" should be included before the phrase "After Action Report/Improvement Plan" on the cover page and in the header/footer of all versions except the final AAR/IP.]

This page is intentionally blank.

Homeland Security Exercise and Evaluation Program (HSEEP)

After Action Report/Improvement Plan (AAR/IP)	[Full Exercise Name] [Exercise Name Continued]

Administrative Handling Instructions

1. The title of this document is [complete and formal title of document].

2. The information gathered in this AAR/IP is classified as [for official use only (FOUO)] and should be handled as sensitive information not to be disclosed. This document should be safeguarded, handled, transmitted, and stored in accordance with appropriate security directives. Reproduction of this document, in whole or in part, without prior approval from [agency] is prohibited.

3. At a minimum, the attached materials will be disseminated only on a need-to-know basis and when unattended, will be stored in a locked container or area offering sufficient protection against theft, compromise, inadvertent access, and unauthorized disclosure.

4. Points of Contact: [List all points of contact below.]

[Federal POC:]

Name
Title
Agency
Street Address
City, State ZIP
xxx-xxx-xxxx (office)
xxx-xxx-xxxx (cell)
e-mail

[Exercise Director:]

Name
Title
Agency
Street Address
City, State ZIP
xxx-xxx-xxxx (office)
xxx-xxx-xxxx (cell)
e-mail

This page is intentionally blank.

Contents

[If an AAR contains graphics, figures, or tables, they should be numbered and listed in the Contents section (e.g., Figure 1, Table 1, etc.)].

This page is intentionally blank.

Executive Summary

[When writing the executive summary, keep in mind that this section may be the only part of the AAR/IP that some people will read. Introduce this section by stating the full name of the exercise and providing a brief overview of the exercise. This brief overview should discuss why the exercise was conducted; the exercise objectives; and what target capabilities list (TCL) capabilities, activities, and scenario(s) were used to achieve those objectives. All of these areas will be discussed in more detail in the subsequent chapters of the AAR/IP. In addition, the executive summary may be used to summarize any high-level observations that cut across multiple capabilities.]

The [agency or jurisdiction] [scenario type] [exercise type] exercise [exercise name] was developed to test [agency or jurisdiction]'s [capability 1], [capability 2], and [capability 3] capabilities. The exercise planning team was composed of numerous and diverse agencies, including [list of agencies participating in planning team]. The exercise planning team discussed [include a brief overview of the major issues encountered, discussed, and resolved during the exercise planning process. Topics to address in this section could include the length of the planning process, the reasoning behind the planning team's choice of objectives to exercise, etc.].

Based on the exercise planning team's deliberations, the following objectives were developed for [exercise name]:

- Objective 1: [Insert 1 sentence description of the exercise objective]

- Objective 2: [Insert 1 sentence description of the exercise objective]

- Objective 3: [Insert 1 sentence description of the exercise objective]

The purpose of this report is to analyze exercise results, identify strengths to be maintained and built upon, identify potential areas for further improvement, and support development of corrective actions.

[In general, the major strengths and primary areas for improvement should be limited to three each to ensure the executive summary is high-level and concise.]

Major Strengths

The major strengths identified during this exercise are as follows:

- [Use complete sentences to describe each major strength.]

- [Additional major strength]

- [Additional major strength]

Primary Areas for Improvement

Throughout the exercise, several opportunities for improvement in [jurisdiction/organization name]'s ability to respond to the incident were identified. The primary areas for improvement, including recommendations, are as follows:

- [Use complete sentences to state each primary area for improvement and its associated key recommendation(s).]

- [Additional key recommendation]

- [Additional key recommendation]

[End this section by describing the overall exercise as successful or unsuccessful, and briefly state the areas in which subsequent exercises conducted by these jurisdictions and/or organizations should focus.]

Section 1: Exercise Overview

[Information in the exercise overview should be "structured data"—written as a list rather than in paragraph form—in order to facilitate preparation of other parts of the AAR/IP, maintain consistency within AAR/IPs, and facilitate the analysis of AAR/IPs for program reporting.]

Exercise Details

Exercise Name

[Insert formal name of exercise, which should match the name in the header.]

Type of Exercise

[Insert the type of exercise as described in Homeland Security Exercise and Evaluation Program (HSEEP) Volume I: HSEEP Overview and Exercise Program Management (e.g., seminar, workshop, drill, game, tabletop, functional exercise, or full-scale exercise.]

Exercise Start Date

[Insert the month, day, and year that the exercise began.]

Exercise End Date

[Insert the month, day, and year that the exercise ended.]

Duration

[Insert the total length of the exercise, in day or hours, as appropriate.]

Location

[Insert all applicable information regarding the specific location of the exercise; including any city, state, federal region, international country, or military installation.]

Sponsor

[Insert the name of the federal agency or agencies that sponsored the exercise, as well as any co-sponsors if applicable. Also list any applicable points of contacts.]

Program

[Insert the name of the program (e.g., Fiscal Year 2007 State Homeland Security Grant Program) from which exercise funding originated.]

Mission

[Insert the appropriate mission areas of the exercise (e.g., prevent, protect, response, and/or recovery).]

Capabilities

[Insert a list of the target capabilities addressed within the exercise.]

Scenario Type

[Name the exercise scenario type (e.g. chemical release).]

Exercise Planning Team Leadership

[The name of each member of the planning team leadership should be listed along with their role in the exercise, organizational affiliation, job title, mailing address, phone number, and e-mail address.]

Participating Organizations

[Insert a list of the individual participating organizations or agencies, including federal, state, tribal, non-governmental organizations (NGOs), local and international agencies, and contract support companies as applicable.]

[Protective Marking]
Homeland Security Exercise and Evaluation Program (HSEEP)

After Action Report/Improvement Plan	**[Full Exercise Name]**
(AAR/IP)	**[Exercise Name Continued]**

Number of Participants

[Insert a list of the total number of each of the following exercise participants, as applicable:

- Players: [#]

- Controllers: [#]

- Evaluators: [#]

- Facilitators: [#]

- Observers: [#]

- Victim role players: [#]

Section 2: Exercise Design Summary

[The exercise design summary is intended to provide a summary of the exercise design process.]

Exercise Purpose and Design

[This section should contain a brief (one-to-two paragraph) summation of why the exercise was conducted and what the exercise participants hoped to learn. It should also include a brief history of how the exercise was organized, designed, funded, etc.]

Exercise Objectives, Capabilities, and Activities

[The purpose of this section is to list exercise objectives and align them with associated capabilities from the target capabilities list (TCL). For each TCL capability there is an exercise evaluation guide (EEG), which lists specific activities that must be performed to demonstrate a capability. In addition to TCL capabilities, the EEG activities relevant to each objective should also be included in this section. Begin this section with the following text.]

Capabilities-based planning allows for exercise planning teams to develop exercise objectives and observe exercise outcomes through a framework of specific action items that were derived from the target capabilities list (TCL). The capabilities listed below form the foundation for the organization of all objectives and observations in this exercise. Additionally, each capability is linked to several corresponding activities and tasks to provide additional detail.

Based upon the identified exercise objectives below, the exercise planning team has decided to demonstrate the following capabilities during this exercise:

- **Objective 1:** [Insert a one-sentence description of each objective].

  – **[Capability Title]**: [Activity 1]; [Activity 2]; and [Activity 3].

  – **[Capability Title]**: [Activity 1]; [Activity 2]; and [Activity 3].

Scenario Summary

[For an operations-based exercise, this section should summarize the scenario or situation initially presented to players, subsequent key events introduced into play, and the time in which these events occurred. For a discussion-based exercise, this section should outline the scenario used and/or modules presented to participants.]

Section 3: Analysis of Capabilities

This section of the report reviews the performance of the exercised capabilities, activities, and tasks. In this section, observations are organized by capability and associated activities. The capabilities linked to the exercise objectives of [full exercise name] are listed below, followed by corresponding activities. Each activity is followed by related observations, which include references, analysis, and recommendations.

[The format for section 3, as described above, represents the preferred order for analysis of exercise observations. However, observations that are cross-cutting and do not apply to one specific activity within the capability should be listed first, directly under the capability summary. Below the cross-cutting observations, you may then present the complete list of activities which apply to the observation.]

Capability 1: [Capability Name]

Capability Summary: [Include a detailed overview of the capability, drawn from the TCL capability description, and a description of how the capability was performed during an operations-based exercise or addressed during a discussion-based exercise. The exact length of this summary will depend on the scope of the exercise.]

Activity 1.1: [Using the EEGs, identify the activity to which the observation(s) below pertain.]

    Observation 1.1: [Begin this section with a heading indicating whether the observation is a "strength" or an "area for improvement." A strength is an observed action, behavior, procedure, and/or practice that is worthy of recognition and special notice. Areas for improvement are those areas in which the evaluator observed that a necessary task was not performed or that a task was performed with notable problems. Following this heading, insert a short, complete sentence that describes the general observation.]

References: [List relevant plans, policies, procedures, laws, and/ or regulations, or sections of these plans, policies, procedures, laws, and/or regulations. If no references apply to the observation, it is acceptable to simply list "N/A" or "not applicable."]

1. [Name of the task and the applicable plans, policies, procedures, laws, and/or regulations, and one or two sentences describing their relation to the task]

2. [Name of the task and the applicable plans, policies, procedures, laws, and/or regulations, and one or two sentences describing their relation to the task]

3. [Name of the task and the applicable plans, policies, procedures, laws, and/or regulations, and one or two sentences describing their relation to the task]

Analysis: [The analysis section should be the most detailed section of section 3. Include a description of the behavior or actions at the core of the observation, as well as a brief description of what happened and the consequence(s) (positive or negative) of the action or behavior. If an action was performed successfully, include any relevant innovative approaches utilized by the exercise participants. If an action was not performed adequately, the root causes contributing to the shortcoming must be identified.]

Recommendations: [Insert recommendations to address identified areas for improvement, based on the judgment and experience of the evaluation team. If the observation was identified as a strength, without corresponding recommendations, insert "none."]

1. [Complete description of recommendation]

2. [Complete description of recommendation]

3. [Complete description of recommendation]

[Continue to add additional observations, references, analyses, and recommendations for each capability as necessary. Maintain numbering convention to allow for easy reference.]

Section 4: Conclusion

[This section is a conclusion for the entire document. It provides an overall summary to the report. It should include the demonstrated capabilities, lessons learned, major recommendations, and a summary of what steps should be taken to ensure that the concluding results will help to further refine plans, policies, procedures, and training for this type of incident.

Subheadings are not necessary and the level of detail in this section does not need to be as comprehensive as that in the executive summary.]

[Protective Marking]
Homeland Security Exercise and Evaluation Program (HSEEP)

After Action Report/Improvement Plan	**[Full Exercise Name]**
(AAR/IP)	**[Exercise Name Continued]**

Appendix A: Improvement Plan

This IP has been developed specifically for [identify the state, county, jurisdiction, etc., as applicable] as a result of [full exercise name] conducted on [date of exercise]. These recommendations draw on both the after action report and the after action conference. [The IP should include the key recommendations and corrective actions identified in Section 3: Analysis of Capabilities, the after action conference, and the EEGs. The IP has been formatted to align with the corrective action program system.]

Homeland Security Exercise and Evaluation Program (HSEEP)

After Action Report/Improvement Plan (AAR/IP)

Table A–1: Improvement plan matrix

Capability	Observation Title	Recommendation	Corrective Action Description	Capability Element	Primary Responsible Agency	Agency POC	Start Date	Completion Date
[Capability 1: Capability name]	1. Observation 1	1.1 Insert recommendation 1	1.1.1 Insert corrective action 1	Planning	State X EMA	EMA director	Mon d, yyyy	Mon d, yy
			1.1.2 Insert corrective action 2	Planning	State X EMS system	EMS system director	Mon d, yyyy	Mon d, yyyy
		1.2 Insert recommendation 2	1.2.1 Insert corrective action 1	Training	State X EMA	EMA director	Mon d, yyyy	Mon d, yyyy
			1.2.2 Insert corrective action 2	Systems/equipment	State X EMA	EMA director	Mon d, yyyy	Mon d, yyyy
	2. Observation 2	2.1 Insert recommendation 1	2.1.1 Insert corrective action 1	Planning	State X EMS system	EMS system director	Mon d, yyyy	Mon d, yyyy
			2.1.2 Insert corrective action 2	Systems/equipment	State X EMA	EMA director	Mon d, yyyy	Mon d, yyyy

Appendix A: Improvement Plan

[Jurisdiction]

After Action Report/Improvement Plan (AAR/IP)	[Full Exercise Name] [Exercise Name Continued]

[Optional]

Appendix B: Lessons Learned

While the after action report/improvement plan includes recommendations that support development of specific post-exercise corrective actions, exercises may also reveal lessons learned that can be shared with the broader homeland security audience. The Department of Homeland Security (DHS) maintains the Lessons Learned Information Sharing (LLIS.gov) system as a means of sharing post-exercise lessons learned with the emergency response community. This appendix provides jurisdictions and organizations with an opportunity to nominate lessons learned from exercises for sharing on LLIS.gov.

For reference, the following are the categories and definitions used in LLIS.gov:

- **Lesson Learned:** Knowledge and experience, positive or negative, derived from actual incidents such as the 9/11 attacks and hurricane Katrina, as well as those derived from observations and historical study of operations, training, and exercises.

- **Best Practices:** Exemplary, peer-validated techniques, procedures, good ideas, or solutions that work and are solidly grounded in actual operations, training, and exercise experience.

- **Good Stories:** Exemplary but non-peer-validated initiatives (implemented by various jurisdictions) that have shown success in their specific environments and that may provide useful information to other communities and organizations.

- **Practice Note:** A brief description of innovative practices, procedures, methods, programs, or tactics that an organization uses to adapt to changing conditions or to overcome an obstacle or challenge.

Exercise Lessons Learned

[Insert an account of any observations nominated for inclusion in the DHS LLIS.gov system. If there are no nominations, a simple statement to that effect should be included here.]

[Optional]

Appendix C: Participant Feedback Summary

[Appendix C of the AAR/IP should provide a summary of the feedback received through this form.]

[Optional]

Appendix D: Exercise Events Summary Table

[In formulating its analysis, the evaluation team may assemble a timeline of key exercise events. While it is not necessary to include this timeline in the main body of the AAR/IP, the evaluation team may find value in including it as an appendix. If so, this section should summarize what actually happened during the exercise in a timeline table format. The focus of this section is on what inputs were actually presented to the players and what actions the players took during the exercise. Successful development of this section is aided by the design, development, and planning actions of the exercise design team. Prior to the exercise, the exercise design team should have developed a timeline of anticipated key events.]

[An example of the format for the exercise events summary table is presented below.]

Table D–1 Exercise events summary

Date	Time	Scenario Event, Simulated Player Inject, Player Action	Event/Action
02/20/06	0900	Scenario event	Explosion and injuries reported at subway station 13
02/20/06	0902	Player action	Subway services stopped in accordance with protocols; notifications started
02/20/06	0915	Player action	Evacuation ordered for planning zone 2A
02/20/06	0940	Simulated player inject	Traffic at a standstill on major egress route 1 reported to players. (Response generated issue because personnel to staff traffic control points were not deployed.)

[Optional]

Appendix E: Performance Rating

[When a jurisdiction/organization elects to use performance ratings, or when initiatives require a rating within the AAR/IP, the following approach can be used. A qualitative performance rating is assigned to each activity demonstrated within its capability area. The performance rating is based on a systemic review by the lead evaluator of exercise performance based on evaluator analysis of how well the participants demonstrated the capability outcome. The results should be summarized within this appendix and should be based on the supporting narrative contained within the body of the AAR/IP.]

The performance rating categories refer to how well each activity was performed during the exercise and are detailed in the table below.

Table E–1 Performance ratings

Rating	Description
Performed without challenges	The performance measures and tasks associated with the activity were completed in a manner that achieved the objective(s) and did not negatively impact the performance of other activities. Performance of this activity did not contribute to additional health and/or safety risks for the public or for emergency workers, and it was conducted in accordance with applicable plans, policies, procedures, regulations, and laws.
Performed with some challenges, but adequately	The performance measures and tasks associated with the activity were completed in a manner that achieved the objective(s) and did not negatively impact the performance of other activities. Performance of this activity did not contribute to additional health and/or safety risks for the public or for emergency workers, and it was conducted in accordance with applicable plans, policies, procedures, regulations, and laws. However, opportunities to enhance effectiveness and/or efficiency were identified.
Performed with major challenges	The performance measures and tasks associated with the activity were completed in a manner that achieved the objective(s), but some or all of the following were observed: demonstrated performance had a negative impact on the performance of other activities; contributed to additional health and/or safety risks for the public or for emergency workers; and/or, was not conducted in accordance with applicable plans, policies, procedures, regulations, and laws.
Unable to be performed	The performance measures and tasks associated with the activity were not performed in a manner that achieved the objective(s).

Appendix F: Acronyms

[Any acronym used in the after action report should be listed alphabetically and spelled out.]

Table F–1 Acronyms

Acronym	Meaning

INDEX

A

advanced disaster life support (ADLS), 92

advanced life support (ALS), 54, 84

aeromedical unit. *See* air transportation

after action conference, 222

after action review (AAR), 217
  distribution of, 222
  exercise evaluation and, 221
  information received from, 221–222
  minimum data elements in, 221

after action review (AAR) template, 225–249
  acronyms used, 249
  administrative handling instructions for, 227
  capability analysis, 238–239
  conclusion, 240
  content list for, 229
  executive summary of, 231–232
  exercise design summary, 236–237

exercise events summary table, 246

exercise overview and, 233–235

improvement areas for, 232

improvement plan, 241

improvement plan matrix, 242

lessons learned, 243–244

participant feedback summary, 245

performance rating, 247–248

scenario description, 237

strengths for, 232

aid
  job, 219
  mutual, 119–121, 186–187
  stations, 185

air operations branch director (AOBD), 21

air transportation, 105–106, 109
  considerations for, 106
  location needed for, 106

alternative strategies assembly, 24

ambulance
  coordinator, 107
  national contract for, 152–153